The Idealized Mind

The Idealized Mind

From Model-Based Science to Cognitive Science

Michael D. Kirchhoff

The MIT Press
Cambridge, Massachusetts
London, England

The MIT Press
Massachusetts Institute of Technology
77 Massachusetts Avenue, Cambridge, MA 02139
mitpress.mit.edu

The MIT Press would like to thank the anonymous peer reviewers who provided comments on drafts of this book. The generous work of academic experts is essential for establishing the authority and quality of our publications. We acknowledge with gratitude the contributions of these otherwise uncredited readers.

This book was set in Stone Serif and Stone Sans by Westchester Publishing Services. Printed and bound in the United States of America.

Library of Congress Cataloging-in-Publication Data

Names: Kirchhoff, Michael D. author
Title: The idealized mind : from model-based science to cognitive science / Michael D. Kirchhoff.
Description: Cambridge, Massachusetts : The MIT Press, [2025] | Includes bibliographical references and index.
Identifiers: LCCN 2024037283 (print) | LCCN 2024037284 (ebook) | ISBN 9780262552936 paperback | ISBN 9780262383943 pdf | ISBN 9780262383950 epub
Subjects: LCSH: Philosophy and cognitive science | Philosophy of mind | Ideals (Philosophy) | Science—Philosophy
Classification: LCC BF311 .K56 2025 (print) | LCC BF311 (ebook) | DDC 153.4—dc23/eng/20250310
LC record available at https://lccn.loc.gov/2024037283
LC ebook record available at https://lccn.loc.gov/2024037284

10 9 8 7 6 5 4 3 2 1

EU Authorised Representative: Easy Access System Europe, Mustamäe tee 50, 10621 Tallinn, Estonia | Email: gpsr.requests@easproject.com

For my family

Contents

Preface ix

Introduction 1

I Scientific Realism and the Nonliteral View of Idealization 21

1 Scientific Models and the Map Problem 23
2 Idealization: The Standard View 37
3 Idealization: The Nonliteral View 47
4 Idealization and Scientific Realism 65

II Implications for Cognitive Science 85

5 Idealization and Neural Representation 97
6 Idealization and Content in Computational Modeling 127
7 Idealization and the Mechanistic Theory of Computation 137
8 Idealization, Multiple Models, and Pluralism 149
9 Idealization and a Grand Unifying Theory of the Brain (Case Study) 171
10 Idealization and Fictionalism in Science and Cognitive Science 197

Concluding Remarks: The Idealized Mind 215

References 217
Index 233

Preface

This book is best seen as sitting at the intersection between philosophy of cognitive science and philosophy of science. Its core message is: scientific models of the mind are *idealized* models. We study nature, including the mind, by partially or entirely idealizing it. An idealized model is a description of a *hypothetical* system—something that does not actually exist in nature. This is the first claim I wish to defend about scientific models in cognitive science. The second claim to be defended is that even if models of the mind are idealizations, this does not stand in the way of defending a version of scientific realism about scientific models in cognitive science.

My interest in discussions about scientific modeling, idealization and scientific realism started with a coauthored paper "The Literalist Fallacy and the Free Energy Principle: Model-building, Scientific Realism and Instrumentalism" accepted for publication in the *British Journal for the Philosophy of Science* (Kirchhoff, Kiverstein, and Robertson forthcoming). The motivation for writing this paper was the observation that several philosophers of cognitive science were inferring the truth of instrumentalism (a version of scientific antirealism) from the fact that theoretical frameworks in neuroscience make use of idealizations. This inference is fallacious; or so we argued. We called it the *literalist fallacy*. The literalist fallacy is the fallacy of accepting or affirming instrumentalism based on the claim that the theoretical constructs of scientific models do not literally map onto real-world, target systems. However, the widespread use of idealization in science does not entail any kind of scientific antirealism. Therefore, even if models of the mind are both simplified and idealized, this does not stand in the way of defending a version of scientific realism in the sciences of the mind. This is the position defended in this book.

To get us this far, I start with discussions about scientific modeling, idealization and scientific realism in philosophy of science. I do not intend this to be a mere overview of how these topics are discussed by philosophers of science. Ultimately, I disagree with how most philosophers of science characterize what idealizations are. The usual account of idealization states that idealizations are distortions of reality or useful falsehoods about target systems. I deny this claim. I argue that some idealizations are useful but that idealizations are neither true nor false with respect to the target systems under investigation. The reason is that idealized posits should not be interpreted as being truth-apt because they do not serve any representational function across the model-world relation. This has real implications for how to think about scientific realism, for how best to defend scientific realism about idealized models in cognitive science, and for how to understand the metaphysical status and standing of core theoretical posits in cognitive science.

Many people have helped me in writing this book. My journey started in 2017, where I was awarded a John Templeton Foundation Academic Cross-Training Fellowship to undertake formal training in theoretical neuroscience at the Wellcome Trust Centre for Neuroimaging at University College London under the mentoring of Karl Friston. I would not have been able to deal with the many technical details of neuroscience and cognitive science without the help and guidance of Prof. Friston. Thanks to my colleague and Head of School, Daniel D. Hutto for enabling me to spend Spring 2023 on research leave. This allowed me to make serious progress on this project. Hutto has been, and continues to be, a source of great intellectual stimulation. Special thanks to Jakob Hohwy and Elena Walsh for generous comments on the introduction of the book—forcing me to more clearly articulate its significance, aim and intended audience. I am especially grateful to Elena Walsh for helping me understand the function of limit functions discussed in chapter 4. Thanks to Anthony Hooper for comments on the final chapter. Thanks also to Julian Kiverstein, Ian Robertson, Erik Myin, Josef Kohlmaier, Sophie Frazer, Ron Planer, Mads Dengsø and Lorena Sganzerla for ongoing conversations about the topics of this book. I am also grateful for extensive comments from three anonymous reviewers of the book manuscript, as well as to two anonymous reviewers of which chapter 9 of the book is based in the *British Journal for the Philosophy of Science*. A special thanks goes to Philip Laughlin and Haley Biermann from the MIT Press, as well as to Madhulika Jain, Mona Tiwary, and Emma Martin for their assistance with the copyediting of the book.

Introduction

We typically understand the world, including the underlying structures of the mind, in an indirect way by building scientific models of the world. Model-based science is a pervasive research strategy, from physics to cognitive science and beyond. A key motivation for using models to explore aspects of reality is the impressive complexity of most, if not all, natural systems. As Williamson notes: "When a system resists direct study, because it is so complex or hard to observe, model-building constitutes a key fallback strategy." (2017, 159; see also Potochnik 2017)

Scientific models have many fascinating features. Two of these will play a central role in this book.

First, most scientific models are unobservable. They can be communicated and represented, but most scientific models cannot be readily observed by looking out into the world. The Hodgkin-Huxley (HH) model, for instance, describes how action potentials are formed and propagated throughout the nervous system (for details, see chapter 1). Action potentials and their characteristics are real parts of nature. However, the HH model itself cannot be readily observed as part of the biological realities of neural dynamics since it is a set of nonlinear differential equations that approximates the electrical properties of excitable neurons. Nonlinear differential equations are part of our scientific models used to understand aspects of reality, even if such equations are not part of the natural systems they are invoked to study. Most scientific models do not exist as part of the fabric of the natural world. A common view has it that scientific models are abstract explanatory devices that exist (mainly) in the imagination of the scientific communities that use them to describe, explain, and predict aspects of the natural world.

Second, scientific models are, almost exclusively and unavoidably, *idealizations* of the world we seek to understand. Idealizations are everywhere in science. Standard examples include: the ideal gas law, frictionless planes, infinite biological populations, neural networks trained on "unrealistic learning rules" (Godfrey-Smith 2009), point objects, state spaces, infinitesimal limits, isolated populations, perfectly rational agents, and representing people as mutually disinterested as it is done in Rawls's original position (1971). The standard view of idealization has it that idealizations are useful and legitimate *falsehoods*—a way of deliberately distorting reality (Weisberg 2013), or a way of habitually lying about some aspects of natural phenomena under investigation (Thompson-Jones 2005). Fletcher et al. articulate the standard view of idealization as follows: "All idealizations . . . involve distortions or *misrepresentation*. It is in this context that we enquire into the coherency and plausibility of appealing to, strictly speaking, *falsities*, in order to discover truth" (2019, 1659–1660; emphasis added). Godfrey-Smith captures both of these peculiar features of scientific models nicely:

> [In model-based science, this] is the style in which a paper may begin: "imagine a population of self-replicating molecules . . . ," "assume a three layer neural network learning by back-propagation . . . ," or "consider a collection of agents playing one-shot prisoner's dilemmas at random . . ." What this seems to be is a style of theoretical work in which *an imaginary system is introduced and investigated*—an imaginary population, ecology, neural network, stock market, or society. The behavior of the imaginary system is explored, and this is used *as the basis for an understanding of more complex real-world systems*. (Godfrey-Smith 2009, 102; emphasis added)

My central aim is to argue that (1) our best models of the mind are idealized models, and (2) despite being both simplified and idealized, such scientific models are still consistent with a version of scientific realism—the view that one actual and reasonable aim of science is to provide true (or, approximately true) descriptions of reality. To arrive at this conclusion, this book seeks to address a novel combination of topics across the philosophy of modeling (and philosophy of science in general) and the philosophy of cognitive science. There is a large literature within the philosophy of cognitive science that focuses on core issues such as neural representation, neural computation and the prospects of explanatory unification. Yet, these discussions hardly mention methodological issues concerning scientific modeling and idealization (and if and where they do, idealization takes a back seat, if it is given a seat at all). This is unfortunate because cognitive science is predominantly a model-based science. Conversely, there is a fast-growing literature on scientific modeling in philosophy of science. However, here one finds no discussion of cognitive science—the focus is mostly on examples from physics. This book rectifies this by bringing these two fields of research together (see Stinson 2020 for some initial attempts at merging the philosophy of modeling with discussions about artificial intelligence in philosophy of cognitive science). My hope is that integrating the literature on idealization and scientific modeling with current work in cognitive science will help seed the ground for a new model-based paradigm for cognitive science and its philosophy: a paradigm where all the central issues about scientific modeling are at the heart of research in cognitive science.

I argue that insights about the importance of idealization in science, and how the rampant use of idealization relates to realism about scientific theories, have direct implications for work in cognitive science, especially concerning the status and standing of neural representation, neural computation, and the prospects of explanatory unification.

To appreciate this, consider what Maley and Piccinini say here: "Computationalism is here to stay. There is every reason to suppose that cognitive capacities have computational explanations, at least in a generic sense. Moreover, everyone agrees with (or should agree with) connectionists or computational neuroscientists, at least in the general sense of embracing neural computation over neural representations. Nonetheless, much work remains to be done" (2014, 92). The quote captures a widespread tendency in cognitive science and philosophy of cognitive science. First is the commitment to the idea that cognition is computational. Because of this, the next step has been to attempt to figure out what an account of computation should look like. In most cases, there is a commitment to the view that computation is a form of information processing over (neural) representations. So, a major issue has been to articulate a defensible account of representation.

Here is the problem. All computational models of the brain and cognitive activity produced so far are either idealizations, making the models hypothetical, or approximations, making the models inexact descriptions of their target phenomena, or a mix of both. Although concepts such as neural representation, neural computation, and information processing strike many as being intimately tied to our understanding of the mind and brain, they may be no more than the idealized posits of computational neuroscience and cognitive science. Their function may be no more than idealizations such as point objects in physics and infinite populations biology. Moreover, it is one thing to get our theories and models technically correct. It is a different issue altogether to know whether scientific theories and models accurately reflect the realities they are invoked to explain. The Ptolemaic model in astronomy is largely internally coherent but ultimately false. Indeed, even if technically correct, there may still be large discrepancies between scientific models and the realities they claim to represent. Consider the ideal gas model. It represents gas molecules as perfectly elastic spheres that occupy negligible space and exhibit no mutual attraction. The ideal gas model is (mathematically) internally coherent. Yet, there are no ideal gases in nature. The same is true of computational models in neuroscience. Consider the hugely influential Bayesian approach to computational neuroscience. Bayesian computational models are a subset of artificial neural networks used as a predictive tool in neuroscience. Bayesian models make assumptions that are known to be false about the brain—for example, that nervous systems update their internal dynamics based on Bayes's rule. Bayesian models are internally coherent. Yet, Bayesian models do not have any concrete existence in the brain—just as ideal gasses have no concrete existence in nature. Here is what follows: One can still explain (or approximately explain) cognition computationally. However, it is a mistake to infer from this that cognitive processes literally are computational processes. Hence, computationalism might be here to stay, but only in the sense that computational tools provide a useful way of modeling the mind and brain.

In realizing the central aims of this book, I therefore have two agendas: the first is an exploration of how best to defend a version of scientific realism in the context of idealized model-based science (part I); the second is an application of this discussion to foundational topics in cognitive science, especially about the status of neural representation, neural computation, and the prospects of explanatory unification (part II).

Part I of the Book

I start by defending a version of scientific realism in idealized model-based science. Scientific realism comes in different shapes and sizes, with differing commitments. Yet, scientific realists generally adopt the attitude that it makes sense to say that science seeks to describe and explain the real structures of the world we inhabit—from physics and beyond. Most scientific realists are also committed to a version of *semantic realism*. This is, broadly speaking, the view that scientific discourse is *truth-conditional*—scientific claims must be either true or false. I argue that *literalism* about scientific discourse (= semantic realism) is not needed to defend a version of scientific realism. Specifically, I argue that scientific discourse about idealizations is best understood as lacking truth-values, when assessed from a model to a target system. The reason for this is that idealizations do not serve any representational function. I call this the *nonliteral view of idealization.* I argue that treating at least some scientific discourse nonliterally is the best way to defend a version of scientific realism that does not give up on truth as the central epistemic aim of science. What's at stake? Saving the widely received view that truth matters to science and its success.

Part II of the Book

My second goal is to explore and investigate the extent to which scientific realism can be defended about some of the most foundational topics in the cognitive sciences: neural representation, neural computation, and the prospects of explanatory unification. A ubiquitous feature of cognitive science is the prominence of semantic concepts such as truth, information, and communication. Equally widespread is talk of codes, algorithms, encoding and decoding, and so on—discourse inherited from computer science and related disciplines. I argue that the most plausible understanding of these informational and computational notions is that they are idealizations. Idealizations might be scientifically useful, but they have no actual existence in the brain. The central thesis is that models of the mind are idealized models—just like most scientific models. We study nature by idealizing it. Infinite populations are introduced in biology to facilitate understanding of phenotypic distribution across generations. Techniques from machine learning are imported into computational neuroscience to solve analytically intractable problems. And so on. In my view, minds are as much a part of nature as natural gases, the planets in our solar system, and the many neurons firing each second in the human nervous system. Therefore, when studying the mind, one should expect the methods to be (more or less) the same as the ones we use to study the rest of the natural world. One of these methods is the construction of idealized models, which one can use to draw coherent (and incoherent) inferences about patterns in reality.

My claim that models of the mind are idealizations is controversial. It is controversial because it forces us to rethink some of the most fundamental constructs used to explain the nature and function of the mind. It is common currency to think that mental activity consists of computational processes over representational vehicles with (semantic) content. The mainstream view is an exemplification of entity realism about *X*s, where entity realism implies the actual existence of *X*s. Entity realism about neural representation is, therefore, the view that representations literally exist in nervous systems. It is in virtue of their objective existence that frameworks in cognitive science that appeal to neural representations are explanatorily successful. Similarly, for commitments to the view that cognitive processes literally are computational processes over neural representations.

I challenge this realist view of the mind. I argue that theoretical constructs such as neural representation and neural computation are located on the side of how we model the structure and function of the mind/brain, not on the side of the nervous system itself. In this sense, both the representational theory of mind and the computational theory of mind are idealizations. This claim is controversial. As I said above, the orthodox view is that minds are both representational and computational systems (realism). Realism is not the only option. It is becoming increasingly fashionable to argue that appeals to the idea that minds are representational and computational systems are merely appeals to useful fictions (fictionalism). Others, still, are seeking to eliminate the use of these constructs from how we model and understand the mind (eliminativism).

In defending the view that our best models of the mind/brain are idealizations, I end up rejecting all these options. I deny realism, fictionalism, and eliminativism. The view I endorse is realism about scientific models that appeal to such theoretical constructs to understand the mechanisms, networks, and causal dynamics of the brain. Idealized models of the mind, or so I argue, are wholly consistent with a version of scientific realism—even for models that include reference to neural representation and neural computation.

Even if it might be controversial to say that models of the mind are idealizations, examples of idealizing the mind are easy to come by. There are a variety of computational models of the mind, from the language of thought hypothesis to Bayesian models of the mind. These are all idealizations. A Bayesian model that describes how an "ideal observer" would estimate environmental conditions is a model of an agent who exists only in the imagination of those using Bayesian models to describe their intended target systems. In this sense, Bayesian models of the mind are idealizations. They are unrealistic in the same sense that postulates about infinite populations and frictionless planes are idealizations in biology and physics (see, e.g., Rescorla 2015). Or, consider that pretty much everyone in the cognitive sciences takes minds to be, almost by definition, processors of information. This is an idealization, since it forces one to think of information as a kind of thing or commodity (see, e.g., Ramsey 2020). Indeed, it might simply be a theoretical confusion to understand information as something that could be processed. That is, if one understands information in terms of covariance—a respectable and scientifically robust notion of information—then information is not something that can in truth be processed, sent, stored, or transmitted. Covariation is a relation. It is not a manipulable commodity (Sganzerla, Hutto, and Kirchhoff forthcoming). There are many more examples, and I will be discussing them as I move across chapters 5–10.

The Argument

My overall argument is formulated in two parts. Part I develops a novel view of idealization in science, which I call the *nonliteral view of idealization*. I use this nonliteral view of idealization to show how scientific realism can be defended in the context of model-based science. In part II, I apply the nonliteral view of idealization to some of the most foundational topics in the philosophy of cognitive science—in particular, neural computation, neural representation, and explanatory unification. I show how appeals to neural representations and neural computation are appeals to idealizations and why this does not stand in the way of endorsing a version of scientific realism about models that include such idealizations. I then go on to show how it is possible to defend the prospects of explanatory unification in cognitive science despite this area of science making heavy use of multiple different and highly idealized models. I end by arguing that the nonliteral view of idealization defended in this book is not any kind of fictionalism.

Part I of the Book in More Detail

Part I concerns my general aim of exploring and defending a version of scientific realism in the context of idealized model-based science (chapters 2–4). The structure of part I is illustrated in figure 0.1.

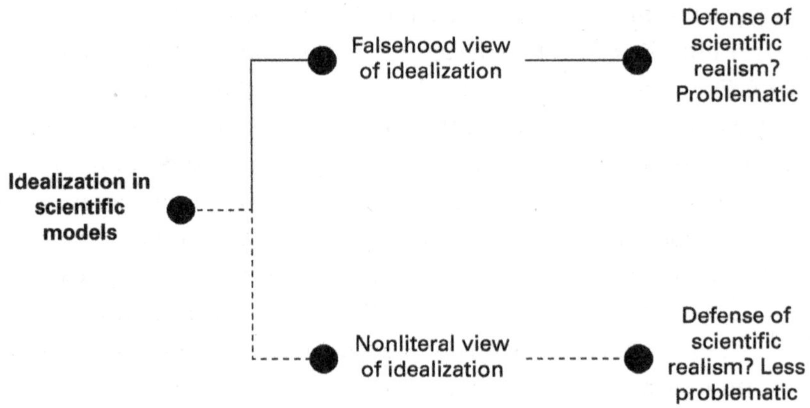

Figure 0.1
The argumentative path of part I of this book. The solid line refers to a route of argument that ultimately makes scientific realism problematic. The dashed line represents what I argue for in this book and highlights the most promising route for a defense of scientific realism.

What I call the *falsehood view of idealization* (FvI) is ubiquitous in contemporary philosophy of science. The FvI asserts that idealization involves the deliberate introduction of a distortion of reality into a model. Hence, the FvI is committed to the idea that idealizations are a way of misrepresenting reality (for certain kinds of explanatory gain). By implication, since idealizations are misrepresentations of reality, they are falsehoods when taken to refer to aspects of reality. I provide several examples of the FvI in chapters 2 and 3. The FvI is based on literalism (or semantic realism). Literalism is the thesis that scientific discourse must be taken at face value. Scientific claims are either true or false. Armed with literalism, idealized posits in science come out, by definition, as false. If taken literally, claims about infinite populations, and so on, misrepresent reality, even if they play a crucial role in the progress of science (chapters 2–3).

Most formulations of scientific realism are based on a *veritist* epistemology—that truth or truthlikeness is a requirement for the epistemic acceptability of scientific claims. The first conclusion I shall make is that the FvI undermines veritism. It undermines veritism because the FvI renders scientific discourse frequently false (chapter 4). Hence, the FvI removes the ground for a feasible defense of scientific realism (chapter 4). I take this conclusion to be both controversial and surprising, especially given that many seeking to show that scientific realism is not threatened by idealization in science are (explicitly) working with the FvI. I shall therefore argue that one should avoid the solid line illustrated in figure 0.1.

I opt to travel along the dashed line in this book by advancing a counter thesis—that idealizations are useful and legitimate scientific practices but that idealizations are not falsehoods. Idealizations are not falsehoods, because the function of idealization is not to represent aspects of reality (chapter 3). I call this the *nonliteral view of idealization.* I maintain that the nonliteral view of idealization is true, and my aim across chapters 3–4 is to make this a plausible position to adopt. There are other nonliteral proposals about scientific discourse in the literature, most notably by Frigg and Nguyen (2021) and Rowbottom (2019, 2022). As I move across chapters 3–4, I address how my proposal connects and departs from these views.

Part II of the Book in More Detail

I now turn to part II of the book. It concerns my specific aim of exploring and investigating the extent to which scientific realism can be defended about some of the most foundational topics in the cognitive sciences. In doing so, I apply the nonliteral view of idealization, from chapters 3–4, to these issues. I am concerned with three separate topics and a particular case study:

- Representation and idealization in cognitive science.
- Computation and idealization in cognitive science.
- Explanatory unification and idealization in cognitive science.
- Grand unifying theories and idealization in cognitive science (case study).

The structure of part II, for the discussion of neural representation, neural computation and idealization in cognitive science, is illustrated in figure 0.2.

Computation and representation are widely regarded as central to the foundations of modern cognitive science, even if their roles are controversial. A widely accepted view is that cognitive processes are representational processes, whereas cognitive activity is the computational process of acquiring, using, and transforming representational content (see, e.g., Sterelny 1991).

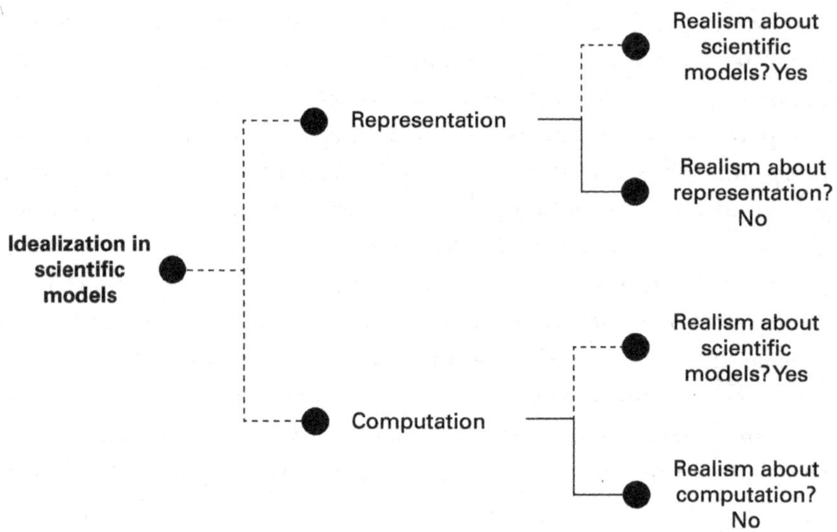

Figure 0.2
The argumentative structure for chapters 5–7 concerning representation and idealization in cognitive science and computation and idealization in cognitive science. The solid line indicates the option one cannot defend. The dashed line refers to routes that are open to a scientific realist defense.

In this book, I treat discussions about computation and representation as special cases of model building in science. I do this because I want to explore how these concepts figure into how we build models of cognitive processes and how (if at all) those models can be turned into truth-conditioned assertions about their respective target phenomena. This is an unusual move in the discussion about neural representation and neural computation (see also Chirimuuta 2024). Most discussions proceed along the lines of assuming that these theoretical constructs have actual existence in target systems (e.g., the nervous system) and then trying to work out the details of an adequate framework for showing both that they exist and why they are explanatorily useful (as we saw in the quote from Maley and Piccinini above). However, by treating computation and representation as cases of model-based science, one must start by acknowledging that computational and representational posits are component parts of simplified and idealized models of the mind. To model the mind, therefore, is either to simplify it or to idealize it, or both.

My working assumption is that there are no substantial differences between models in sciences such as physics and biology and models in cognitive science. From a modeling perspective, there is nothing special about brains and minds—their complexity might be special, but this does not set minds and their neuronal hardware apart from anything else studied in nature. My strategy in chapters 5–7 is to map out the ways in which computational and representational models of the mind are like simplified and idealized models in science. I draw various conclusions. However, for each of the domains of discussion, there is one conclusion that remains the same: computational and representational models of the mind are idealizations. They are, in this sense, not part of the fabric of the realities they are invoked to explain and understand. Computational and representational models of the mind are abstract explanatory devices that exist (mainly) in the imagination of the scientific communities that use them to describe, explain, and predict aspects of the world.

My first conclusion (chapter 5) is: a close look at models of neural representation, formulated based on the representational theory of mind, reveals that the ways in which the appeal to neural representations is used are akin to how idealizations are used in the context of scientific modeling. Therefore, neural representations, or the kind of properties usually associated with representations—such as causal efficacy, truth-evaluability, content determinacy, and so on—function as idealizations. This invites two further conclusions (see figure 0.2). The first is that one can defend scientific realism about scientific models that involve representational posits. One can do this because idealization does not undermine scientific realism (following on from chapters 3–4). However, one cannot defend any kind of entity realism about the existence of neural representations in virtue of their status as idealizations.

One philosophically interesting question is: Are idealized posits such as mental states and informational content functionally analogous to idealizations in physics, such as ideal gases? I argue that idealized posits in science, such as the ideal gas model, and idealized posits in cognitive science are disanalogous. Even if we grant that appeals to neural representations are appeals to idealized posits, mental or neural representational posits do not provide the same explanatory gain as idealizations in areas such as physics.

Most cases of successful idealization in science involve particular systems—e.g., natural gases—where the scientists working with some model—such as the ideal gas model—have a very solid understanding of the parts involved in the model, their respective functions and properties, and how precisely the theoretical model relates to its target system in the world. This is what allows scientists using the ideal gas model to deliberately treat the value of intermolecular interactions as zero. The ideal gas model allows one to interconvert between volumes and molar amounts in chemical equations. A different example is the use of point objects in physics. Say we want to compute the speed and trajectory of a baseball. We could try to add all the relevant details—wind resistance, air resistance, the earth's rotation, and so on. However, trying to include all of these things will make the analysis hopelessly complex. A more promising way is to simplify the problem. We can do this by neglecting air resistance, which is done by modeling the ball as moving in a vacuum. We can also neglect the size and shape of the ball by modeling it as a point object (i.e., as without mass and dimension). These ways of modifying the situation are ways of idealizing what we ultimately want to explain and understand—the speed and trajectory of the baseball. Using idealizations in this case presents us with a problem that is simple enough to calculate (Young and Freedman 2007).

One reason for the disanalogy is that idealization in science is recognized to have distinct functions. They flag that certain variables are causally irrelevant to explanation (this is the case in the ideal gas model); they sharpen our focus on specific causal factors rather than others (this is also the case in the ideal gas model); and they facilitate computational tractability (which is the case in the baseball example). What about mental states? Or, what about appeals to internal representational content? I argue that discourse about neural representation does not facilitate any of these explanatory functions. Ultimately, the real value of idealization in cognitive science is that it gives us reason to treat talk of an inner world of contentful representations as merely metaphorical—as a means of acknowledging that we cannot literally mean what we are saying when making use of these theoretical terms.

The focus of chapters 6 and 7 is on computational theories of cognition. Following Shagrir (2022), it will be useful to distinguish between two senses of the term "computation" in the computational sciences. First, computation may be taken to refer to the pervasive use of computational models, simulations, and other methods for studying certain systems and their respective functions. The field of computational astrophysics is a branch of science where computational tools are used to study the heavens. Second, a different use of the term "computation" has it that one might justifiably use computational techniques to study certain target systems because those systems are computational systems. The second use of the term is often the one we find in computational neuroscience and philosophy of cognitive science, where the nervous system itself is described as a system that engages in computational processes.

My focus is on the second meaning of computation. Consider, for example, what Sun says of work in computational neuroscience: "[Computational] models embody descriptions of cognition in computer algorithms and programs, based on or inspired by artificial intelligence and computer science . . . they impute computational processes onto cognitive functions, and thereby they produce runnable programs" (2023, 4).

Given the complexity of nervous systems, computational models of neural behavior are, almost unavoidably, simplified and idealized models. How do computational models that trade in simplification and idealization relate to their intended target systems? I argue that the assertion that the nature of the brain and cognition is computational is an idealization. Specifically, I argue that the *computational theory of cognition* (CTC) is an idealization insofar as it states that the nature of the brain and cognition is computational in the specific sense displayed by computational modeling.

Traditionally, the CTC has been articulated in terms of the idea that cognition is the manipulation of language-like representations, which, so its defenders contend, enables one to explain productive and systematic cognitive processes. According to classicism, as it is often referred to, cognition involves computations over internal symbolic sentences written in the form of a language of thought (Fodor 1975). The *language of thought* (LOT) hypothesis is an idealization. We can imagine that internal states have a kind of natural language-like structure, even if they do not. Moreover, the LOT requires that computation is digital, that is, with discrete mappings across states and transitions between states. The brain, however, is not a digital computer (Maley 2023; Piccinini 2020). At a sufficiently high level of abstraction one can describe cognitive processes in accordance with the LOT, even if the nervous system does not exhibit the kind of architecture required to implement the LOT. Hence, the LOT is an idealization.

According to the CTC, cognitive processes are explained by internal computations, which in biological organisms are realized by nervous systems. Following Piccinini (2020), we can note that in its general form, the CTC states that neurocognitive processes are computations (see also Chalmers 2011).

What makes a neurocognitive process computational? There are no easy answers to this, and there are several overlapping and mutually inconsistent theories of the CTC in the literature. In this book, I consider two versions of the CTC: Egan's (2018) theory of mathematical content involved in computation and Piccinini's (2020) mechanistic theory of computation. The core feature that makes the CTC an idealization, for both Egan and Piccinini, and basically for everyone working with models of the mind based on the CTC, is the following: state transitions from input, possible internal states, to output are mapped in accordance with a rule. Central questions here are: What are the relevant representations of input and output, and what rules or algorithms govern the input-output transformations or state transitions? I give some specific examples of this across chapters 6–9.

Here is the point I want to make: If this claim is taken literally—that input-output transformations are governed by specific rules or algorithms—then it is false about the target system, the nervous system. The main reason is that the rule-governing aspect of the CTC need not be encoded in the nervous system. If it is not encoded in the nervous system, then it makes no sense to literally describe computations in the nervous system as governed by certain rules or algorithms. A very simple example brings this out. We can model the trajectories of planets in our solar system in accordance with certain laws of motion. Yet, planets do not move the way that they do because they encode those laws. Where are the laws? They are part of our models of the world. Where are the rule-governing aspects of the CTC? They are on the side of the computational models used to investigate state transitions in the nervous system. In this specific sense, the relevant mapping rule is where all other scientific idealizations are: on the side of the scientific model. Idealizations are not in the realities of the phenomena being investigated. This is the case for infinite populations, frictionless planes, and wholly rational economic agents. It is also the case for the rule-governing aspect of the CTC.

Here is how I propose to defend scientific realism about the CTC. I do so by treating scientific discourse about certain aspects of the CTC nonliterally.

In chapter 6, I focus on Egan's (2018) theory of mathematical content for computation. I argue that the mathematical content that specifies a computational model does so only with respect to the computational model, not the nervous system itself. Crucially, I show that if the mathematical content is taken to literally specify the target system under investigation, it idealizes its target domain. I argue that if the mathematical content theory of computation is treated in terms of the FvI, then it is false. However, if treated in terms of the nonliteral view of idealization, then Egan's proposal can be made consistent with a version of scientific realism. Hence, in chapter 6, I showcase a way of justifying scientific realism about computational explanation without this implying realism about the brain literally doing computational work.

In chapter 7, I turn to discuss Piccinini's (2020) influential mechanistic theory of computation. Here I draw several conclusions. Piccinini (2020), like most, takes computations to be done in accordance with rules. This is true. However, it is true only on the side of the scientific model. If taken literally, it is false of the nervous system. It is worth clarifying this in a bit more detail. I argue that idealizations in model-based science should be treated nonliterally, in the sense of being neither true nor false when compared across the model-world relation. It is, however, also possible to treat the rule-governing condition on computation literally and in a way that would be true. For example, it may be true of a model that its state transitions are computed in accordance with a rule (this much would be true of all computations in a standard machine learning textbook). For example, it may very well be true that a model of edge detection in early vision is computed by taking the Laplacian of a Gaussian distribution (for details, see chapter 6). However, if this claim is taken literally from the model to the target system, it is false. My suggestion is that to save scientific realism about computational explanation in cognitive science, computational discourse must be treated nonliterally. However, given that the rule-governing element is part of the definition of what computation is, I conclude that entity realism about neural computation is false.

The structure of part II, for the discussion of explanatory unification and idealization in cognitive science, can be illustrated in figure 0.3.

Model-based science often makes use of multiple different and often inconsistent idealized models to understand one and the same target system. The basic idea is that a full understanding of complex phenomena requires the use of more than a single model. Weisberg (2007) calls this *multiple-models idealization*. Multiple-models idealization implies explanatory pluralism. Discussions about representational theories of mind and

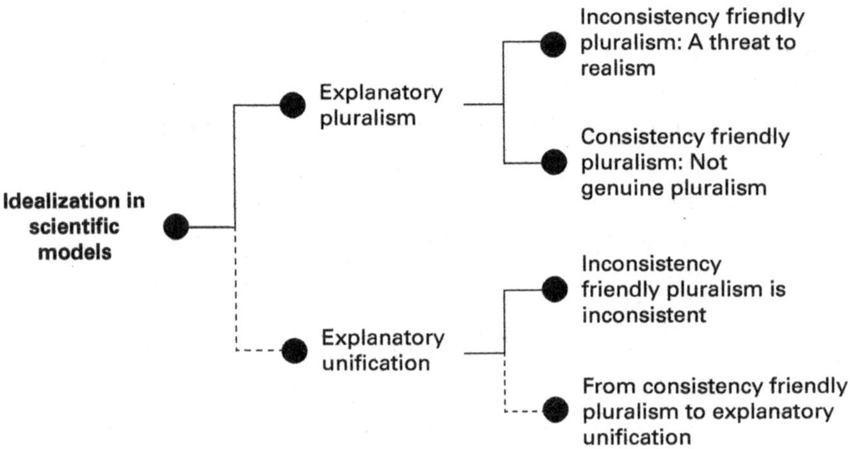

Figure 0.3
The argumentative structure of chapter 8 concerning explanatory unification and idealization in cognitive science. The solid line either makes scientific realism controversial or it renders pluralist proposals not genuinely pluralistic. The dashed gray line represents the perspective I aim to defend, in the sense that it highlights a promising route by which to defend (1) scientific realism and (2) explanatory unification in cognitive science.

computational theories of cognition, however, often proceed on the assumption that we will one day land on a single, unifying theory of cognition that is representational or computational, or some combination of both. Explanatory unification is the name of the game in these discussions. Today, many philosophers of cognitive science are taking an interest in defending explanatory pluralism. I give some concrete examples of this in chapter 8. Some think that representational and nonrepresentational models of cognition can be fruitfully combined. They can. That is, it is not a problem to combine different aspects of different theories into a single model of the mind. Yet, the central question is: Can they be combined in a way that accurately captures the real structures of the mind? What we need to pay attention to is whether different proposals carry with them inconsistent ontological commitments about the basis of mentality (see Hochstein 2023). In chapter 8, I argue that representational and nonrepresentational models of the mind cannot be consistently integrated. They carry with them mutually exclusive ontological commitments. I call pluralism about mutually inconsistent explanatory models "inconsistency-friendly pluralism."

Inconsistency-friendly pluralism gives rise to the well-known "problem of inconsistent models" in the philosophy of science. If multiple-models idealization is treated literally, then the problem of inconsistent models arises. The problem of inconsistent models makes scientific realism hard to defend, since scientific realism cannot be defended if two or more models, all with inconsistent properties, are applied to one and the same part of the world (figure 0.3, top solid line).

Scientific modeling that proceeds in accordance with multiple-models idealization is now being used to tether explanatory pluralism to various forms of scientific realism—in particular, *deflationary* and *perspectival* versions of scientific realism. I call this species of explanatory pluralism "consistency-friendly pluralism." I argue that consistency-friendly pluralism is not any kind of pluralism at all (figure 0.3, second solid line from the top).

In chapter 8, I defend the dashed lines in figure 0.3. In doing so, I aim to sketch a picture of what makes explanatory unification a possible view to defend in cognitive science. I am not saying that this conclusion necessarily follows. Rather, I aim to promote the possibility of explanatory unification. The central line of argument will be a familiar one by now: once we allow that at least some scientific discourse can be taken nonliterally, the sting of the problem of inconsistent models to scientific realism can be defused. This opens up the possibility of starting out with many different models in some domain of inquiry, with the aim of working toward explanatory unification of these models under a single, unifying model of the mind.

The structure of part II, for the discussion about grand unifying theories of cognition and idealization, can be illustrated as in figure 0.4.

In part I, I pin my colors to a particular mast: the view that idealization in model-based science is a useful and legitimate scientific practice but not a falsehood. I call this the nonliteral view of idealization. In chapter 9, I nail my colors to a second mast. I argue that the *free energy principle* (FEP) in theoretical neuroscience is a plausible candidate by which to defend all the different claims made across chapters 1–8.

The FEP is known as a *grand unifying theory* (GUT). It aims to provide a general theory unifying life and mind formulated entirely from mathematical principles in physics and machine learning. The FEP is accompanied by Bayesian mechanics, which provides a Bayesian interpretation of the nature and function of the brain and cognition. Indeed, it provides a Bayesian

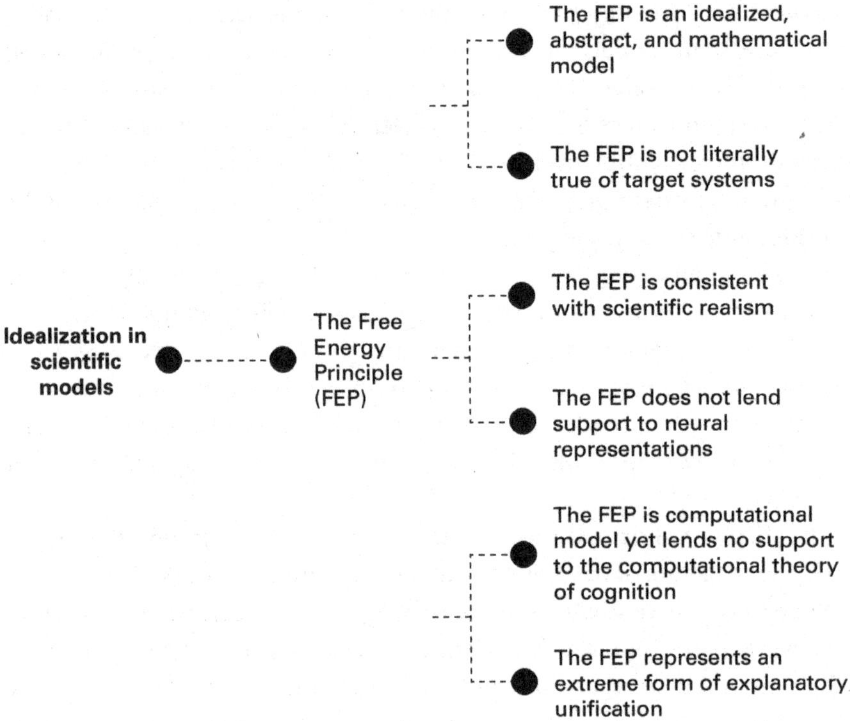

Figure 0.4
The argumentative structure of chapter 9 concerning idealization and grand unifying theories of the brain.

window on the dynamics of self-organization in life. As a GUT, the FEP does not aim to replace other frameworks such as decision theory or reinforcement learning (Parr, Pezzullo, and Friston 2022). Nor does it seek to replace philosophical positions such as enactivism or ecological psychology. Rather, it aims to subsume all these different approaches within one and the same unifying framework (Friston 2010; Kirchhoff 2018a; Kirchhoff et al. 2018; Mann, Pain, and Kirchhoff 2022; Pain, Kirchhoff, and Mann 2022; Robertson, Kiverstein, and Kirchhoff 2024).

I treat the FEP as a case study—one that not only attempts to theoretically unify our thinking about life and mind but also one that integrates all the argumentative threads in this book. In accordance with figure 0.4, I defend the following claims about the FEP:

- The FEP is an idealized, abstract and mathematical model (following on from chapter 1).
- The FEP is not literally true of its intended target domain (following on from chapters 2–3).
- The FEP is consistent with a version of scientific realism based on the nonliteral view of idealization (following on from chapters 3–4).
- The FEP does not lend support to the representational theory of mind (following on from chapter 5).
- The FEP is a computational model yet lends no support to the computational theory of cognition (following on from chapters 6–7).
- The FEP represents an extreme form of explanatory unification (following on from chapter 8).

I said above that I treat the FEP as a case study. Yet, if what I have to say about the FEP, and in terms of how the FEP highlights the key claims across chapters 1–8, then it is a case study with significant implications for modeling-based approaches to cognitive science more generally.

I end this book with a discussion of why the nonliteral view of idealization does not imply any kind of fictionalism about scientific discourse. Its structure takes the form shown in figure 0.5.

Eklund (2024) says that fictionalism "about a region of discourse can provisionally be characterized as the view that claims made within that discourse are not best seen as aiming at literal truth but are better regarded as a sort of 'fiction.'" As a topic of discussion, fictionalism both has a venerable history in philosophy and is currently the subject of intense discussion across almost all areas of philosophy, including the philosophy of cognitive science and the philosophy of science.

The apparent links between the nonliteral view of idealization and fictionalism should be clear. The nonliteral view of idealization denies that claims involving idealizations should be taken literally. However, my conclusion is that the nonliteral view of idealization does not equate to fictionalism (figure 0.5, first solid line). The first argument in support of this claim is that fictionalism about idealizations makes scientific realism problematic (figure 0.5, top solid line). The second argument in support of this claim is that the way in which many discuss fictionalism in the philosophy of science and the philosophy of cognitive science treats fictional posits in

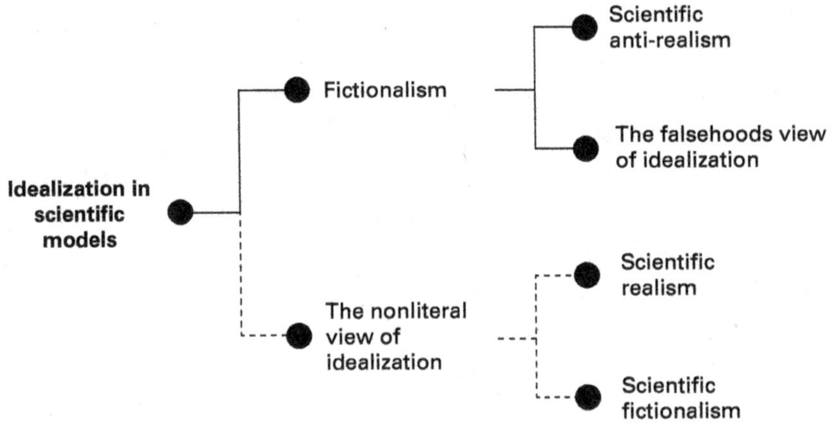

Figure 0.5
The argumentative structure of chapter 10 concerning fictionalism and idealization, covering all of the book. Some equate fictionalism with idealization. This is a mistake (first solid line). Fictionalism, as I shall present it, is inconsistent with scientific realism (top solid line) and rests on the falsehoods view of idealization (second solid line from the top). The nonliteral view of idealization is not any kind of fictionalism. It is a form of scientific fictionalism, which is consistent with scientific realism (dashed lines).

the same way that idealization is treated by those working under the FvI. I argue in chapters 3–9 that the FvI is problematic.

My positive line of argument is represented by the dashed line. Idealizations in model-based science should be understood nonliterally (figure 0.5, first dashed line). The nonliteral view of idealization is the best way of defending scientific realism, as I argue across several chapters (figure 0.5, top dashed line). Finally, I think that the nonliteral view of idealization is a form of scientific fictionalism—the view that the creative imagination is essential to science (figure 0.5, bottom dashed line). Scientific fictionalism is not the same as fictionalism. Once the nonliteral view of idealization is combined with scientific fictionalism, and not fictionalism, the doors are open for a defense of scientific realism—in science, in general, and in the sciences of the mind, in particular.

I Scientific Realism and the Nonliteral View of Idealization

1 Scientific Models and the Map Problem

1.1 Introduction

In science, it is usually the case that a pattern of data cannot be given a description or explanation unless one has a model of the underlying process generating the observed pattern of data (Farrell and Lewandowsky 2018). For instance, it is arguably not possible to explain the apparent retrograde motion of planets in the absence of a model. We need the model to understand the pattern of data we are confronted with. In population biology, Levins (1966) put his finger on the problem of trying to model even simple populations in complete detail. Seeking to capture all the aspects of some part of reality, Levins argued, will result in models being intractable and often entirely incomprehensible to the human modeler (see also Walmsley 2021; Weisberg 2006). The process of modeling complex target systems (i.e., complex phenomena that we want to understand) will (usually) be one of simplification and idealization. In a similar vein, neuroscientists use models to deal with tractability issues when confronted with terabytes of digital output generated by brain scans (Pain, Kirchhoff, and Mann 2022). For this reason, work in computational neuroscience involves the use of models that significantly reduce the number of possible neuronal responses that need to be considered (Dayan and Abbott 2001; Mann, Pain, and Kirchhoff 2022).

An interesting fact about models is that at least some that explain the data will differ substantially from the phenomena they are invoked to explain. For example, the Hodgkin-Huxley (HH) model of the action potential represents different types of ion currents, all of which are said to be explanatorily relevant for understanding the flow of those ions through the cell membrane.

Yet, the HH model does not have ion channels. Ionic channels are proteins. The HH model is a set of equations (Brette 2013).

Or consider that over the last fifty years or so there have been at least three central frameworks for understanding the nature and function of the mind. Following Eliasmith (2003), we can call these *symbolicism*, *connectionism*, and *dynamicism*. Most famously, symbolicism relies on an understanding of the mind as a computer—the mind is the software, the brain is the hardware. Connectionism relies on a different framing of the mind-brain relationship—namely, that the mind is the brain. Dynamicists portray the mind/brain as a control system. Its most famous articulation is found in the work of Van Gelder (1995), where the mind is taken to be a dynamical system exemplified by reference to the Watt governor. Here is what matters. For symbolicists, the canonical properties of the mind are the properties of a Turing machine. A Turing machine is a theoretical computing device developed by Alan Turing (1937) that represents an idealized model for a given mathematical calculation. For connectionists, state transitions—leading to cognitive outcomes—are modeled as real-valued vectors in a high-dimensional state space. A state space is a mathematical representation of the set of all possible configurations of some system. Hence, the use of high-dimensional state spaces is itself both a simplification and an idealization. Finally, for dynamicists, by using the Watt governor as a model of the mind, it became possible to describe mentality—in the language of dynamical systems theory—without appeal to concepts such as representation and computation. Of course, the mind/brain is no Watt governor. If taken literally, the claim that minds are like a Watt governor would be false.

One immediate question one might ask is: If scientific models are invoked to explain aspects of target systems in the world, and if such models deviate from the details of their target systems (often dramatically so), then what is the relation between models and the realities they are invoked to represent? This is in many ways the central question addressed in this book. It concerns the issue of whether or not models, if different from their target systems, are able to accurately or truthfully capture what we want to know about the target system. Yet, before we address this issue, we have some preparatory ground to cover.

The question that shall entertain us in this chapter is: What, precisely, is a scientific model? My central aim is to introduce the reader to several key issues in the literature on scientific modeling. I start by giving some examples of concrete, computational, and mathematical models in science (the distinction between concrete, computational, and mathematical models comes from Weisberg 2013). I finish by raising and considering a few questions one can ask about scientific models and their relation to real-world target systems. The reader should be warned that I do not aim for originality in this chapter. Its sole function is to introduce concepts, distinctions, and other issues that are relevant for the rest of the book.

1.2 Examples of Models

Scientists, whose business is understanding the empirical world, invest a lot of time in both constructing and considering models of the world (Godfrey-Smith 2009). Let us therefore begin by considering some examples of model building in science.

1.2.1 The Hodgkin-Huxley (HH) Model

The HH model of the action potential is one of the greatest success stories in modern neurobiology (Häusser 2020). Koch (1997) agrees, saying that "their quantitative model . . . represents one of the high points of cellular biophysics" (140), culminating in the 1963 Nobel Prize. The HH model consists of a set of differential equations that describe neuronal firing, or *spikes*. An action potential is a sharp and short-lived rise in the electrical potential on the membrane of an axon. The HH model describes the neuronal membrane as an electrical circuit, which involves a capacitor and three parallel series of batteries and variable resistors.

Based on the electrical circuit model of the action potential, the HH model demonstrates three key facts. (1) An action potential arises from the membrane's changing its conductance to particular ions, especially sodium (Na) and potassium (K), thereby generating the action potential. (2) Any change in membrane conductance is conditioned on the membrane's potential. (3) It is possible to intervene on Na and K independently of one another. The quantitative model can be derived from what is known as the total current equation (in Craver 2008, 1023):

$$I_{total} = CdV_m/dt + m\ G_k n^4(V_m - E_{na}) + G_{na} m^3 h(V_m - E_k) + G_1(V_1 - E_1)$$

The first term is a "capacitative" current, which depends on the capacitance of the membrane (C) and the charge voltage (V_m) over time (t). The particular ion currents, (Na) and (K), are modeled in terms of ($I = V/R$), which expresses Ohm's law. Ohm's law states that the voltage across a conductor is proportional to the current flowing through it (assuming that all physical conditions and temperatures remain the same or constant). G_{na} and G_k denote that maximum conductance of the membrane to each ion. ($V_m - E$) refers to the difference between the membrane voltage and the equilibrium potential. Finally, the coefficients n^4 and m^3h are used as constant quantities when fitting the curve for conductance changes based on experimental interventions, which were done with a voltage clamp. In summary, the "total current equation can be used to predict numerous electrophysiological phenomena in the squid giant axon and in neurons generally" (Craver 2008, 1023). This is precisely what Hodgkin and Huxley (1952) did.

1.2.3 The Schelling Model

Is it possible for a city to show evidence of residential segregation along racial lines even if none of the members of the population has any strong preferences for living in racially segregated neighborhoods? The Schelling model of residential segregation is one of the earliest agent-based models seeking to provide a possible answer to this question in social science (Schelling 1978). In the model, an agent occupies a cell of rectangular shape. Only one agent can occupy a cell at any given time. Agents belong to one of two groups and can relocate according to the fraction of *friends* (that is, agents specified as belonging to their own group) within a Moore neighborhood around their location.

Moore neighborhoods are typically used in the construction of cellular automata. Cellular automata are discrete, abstract computational models and are often used as general models of complexity and in specific models of nonlinear dynamics. A Moore neighborhood is usually defined on a two-dimensional square lattice and is typically composed of a central cell and eight cells around it. A Moore neighborhood allows one to model state transitions in complex systems in accordance with relatively simple rules (or algorithms) that determine the state transitions across the two-dimensional lattice.

In the Schelling model, the agents are initially distributed randomly. The agents move based on a utility function and a movement rule: an agent, located in the middle of a Moore neighborhood, where the fraction of friends f is less than a predefined tolerance threshold F (i.e., $f<F$), will

make an effort to move to a different neighborhood, where the fraction of friends will be at least f (i.e., $f \geq F$) (Schelling 1978, 148). According to Weisberg (2012), the "utility function [says] that each individual prefers that at least 30% of its neighbors be of the same type. So the As want at least 30% of their neighbors to be As and likewise for the Bs" (785). Concerning the results of the Schelling model, Weisberg says:

> Schelling's major result is that small preferences for similarity can lead to massive segregation. This result is quite robust across many changes to the model, including different utility functions, different rules for updating, differing neighborhood sizes, and different spatial configurations. In fact, it is extremely hard to avoid segregation when agents have some preference for like neighbors. (Weisberg 2012, 786)

1.2.4 The McCulloch and Pitts Model

In 1943, McCulloch and Pitts published their influential paper "A Logical Calculus of the Ideas Immanent in Nervous Activity." In it, they sought to understand how the brain could generate complex patterns of activity by "simply" using many basic brain cells (neurons) that are connected together. Piccinini (2020), like many others, says of McCulloch and Pitts's work that it presents the first computational theory of the brain and cognition—the aptly named McCulloch and Pitts model. It is arguably true that the McCulloch and Pitts model presents us with the first computational theory of the brain and cognition. However, it is more accurate to say that the work of McCulloch and Pitts (1943) represents the *first heavily idealized and simplified* model for a computational theory of the brain and cognition. In formulating their model, McCulloch and Pitts made the following five assumptions (quoted in Piccinini 2020, 111):

1. The activity of neurons is an "all-or-none" process.
2. A certain fixed number of synapses must be excited within the period of latent addition in order to excite a neuron at any time, and this number is independent of previous activity and position of the neuron.
3. The only significant delay within the nervous system is synaptic delay.
4. The activity of any inhibitory synapses absolutely prevents excitation of the neuron at that time.
5. The structure of the net does not change with time. (McCulloch and Pitts 1943, 101)

All five assumptions are either simplifications or idealizations, or both. Assumption (1) expresses the all-or-none law. That is, a neuron is either at rest or pulsing. According to McCulloch and Pitts, the all-or-none law of nervous activity allows one to model the activity of a neuron in terms of a proposition. In combination with assumption (3), which describes patterns of neuronal activity in terms of *discrete functions*, assumption (1) enabled McCulloch and Pitts to represent state transitions in the nervous system in terms of discrete logical gates: the pattern of neuronal impulses at one state depends on the neuronal impulses of the neuron at a temporally prior state.

What we get from McCulloch and Pitts is a mathematical model of rule-based state transitions of neuronal activity—the kind of activity that is associated with computational processes. That it is a heavily idealized model is beyond question. Abraham notes:

> The McCulloch-Pitts neurons were idealized, and had little detail about how they functioned, which could actually be shown to be more continuous than discrete. They had ignored physiological, empirical details about how neurons function—in particular, the chemical aspect of transmission based on neurotransmitters, in which continuous variables (e.g. chemical concentrations and electrical fields) play a vital role. . . . The McCulloch-Pitts neuron was impoverished: it was assumed for the purposes of modeling to be an on–off device. Yet this formal modeling method held the potential, especially for McCulloch, of making sense of the brain's functional organization and phenomena such as memory, perception, and communication. (Abraham 2018, 56)

In this sense, the McCulloch and Pitts (1943) model represents one of the first computational models of the mind/brain, even if the formal neurons used in the model "were deliberately as impoverished [far removed in their details from actual biological neurons] as possible" (McCulloch 1974, 36; quoted in Piccinini 2020, 112).

1.2.5 The SS *Monterey*

Nguyen and Frigg (2022) ask us to imagine that we are shipbuilders working with ocean liners like the SS *Monterey*. The SS *Monterey* has been tasked with making frequent trips between San Francisco and Honolulu. However, it is too slow from the perspective of the company. Our task is to redesign its engine so that it will be able to sail at a certain average speed while carrying a certain load. The central question becomes one of determining what the minimal power outlet of the engine must be to facilitate its desired function. To calculate this, you will need to construct a scaled-down model of the engine on the actual ship. According to Nguyen and Frigg:

> The model must be carefully constructed: it has to have a shape that reflects the shape of the full-sized ship you're ultimately interested in. And the experiments that you perform on the model have to be carefully designed. In this case, you want to measure the complete resistance, R_c, faced by the model ship as it is propelled through the basin at velocity V, because this gives you information about how powerful the engine needs to be. (Nguyen and Frigg 2022, 1)

1.3 Types of Models: Concrete, Computational, and Mathematical

It is standard practice in the literature to emphasize that current work in model-based science makes use of at least three different *kinds* of models: computational models, concrete models, and mathematical models (and perhaps some mix of these three). The question of interest here is: What is a scientific model?

The ship model in section 1.2.5 is the most straightforward of these three types of scientific models. The ship model is a *concrete material object*. It is a small-scale model of an actual ship in a towing tank. It is not imaginary. It is not a mere possibility. It is an actual, physical object intended to stand in a representational relationship to some real or imagined system—in this case, the SS *Monterey*. Concrete models need not be constructed. One can work with naturally occurring systems in the wild as models for some intended purpose. For example, in the philosophy of mind, honeybee dances are often invoked as a model by which to give a naturalistic explanation of representational content (see, e.g., Neander 2017 for discussion).

The Schelling model is a computational model. So is the McCulloch and Pitts model. A computational model will usually involve a set of procedures that can potentially stand in relation to a computational description of the behavior of the system in the world (see Weisberg 2013). Conway's Game of Life is a different example (Gardner 1970). It seeks to shed light on several difficult questions: When is a physical system alive? When is a physical system cognitive? These questions, as Beer notes, "are so fundamental that practicing biologists, neuroscientists, and cognitive scientists rarely ask them, let alone try to answer them" (2004, 309). In this model, life is modeled as a two-dimensional binary cellular automaton, where the next state of each cell depends only on its own state and the sum of all the other states that comprise the Moore neighborhood in which life is represented. The computational rules that govern the state transitions of a cell are simple: (1) the birth rule, which asserts that an empty, or "dead," cell with exactly three "live" neighbors becomes alive; (2) the death rule, which states that a live cell with zero or one neighbors dies of isolation, and a live cell with four or more neighbors dies of overcrowding; and (3) the survival rule, which says that a live cell with two or three neighbors remains alive. As the game proceeds, across a number of iterations, some cells will remain alive, others will die, and "life-forms" will start to evolve. The Game of Life motivated the use of cellular automata in the field of complexity science, where one finds simulations of such disparate phenomena as ant foraging, traffic scenarios, cloud formation, and galaxies. The Schelling model also makes use of this computational structure to arrive at its outcomes about residential segregation.

The HH model is a mathematical model represented by a set of differential equations. Unlike the ship model, mathematical models are described in the literature as *nonconcrete* or *abstract* models in the sense that they are not material objects per se. Mathematical models are abstract structures whose features may potentially stand in a relation to a mathematical representation of certain phenomena (Weisberg 2013). A different example is the rise of accounting for communication and signaling in biological systems by appealing to information-theoretic notions such as Shannon entropy and mutual information (Planer and Godfrey-Smith 2021). These information-theoretic notions are frequently invoked in theoretical neuroscience when seeking answers to the following kind of question: How much does the neural response tell us about a stimulus? Intuitively, the lower the Shannon entropy is between a neural response and a given (environmental) stimulus, the higher the mutual information between the two—that is, the more information the response carries about the stimulus, and vice versa (see Dayan and Abbott 2001 for the relevant mathematical formalisms). I return to discuss a concrete case of mathematical models used in theoretical neuroscience in chapters 7–9.

1.4 Questions about Models

In addition to there being different kinds of models in science, there are different kinds of questions one can ask about models. Most obviously are what Nguyen and Frigg (2022) call the "semantic question" and the "accuracy question." The semantic question asks us to consider in virtue of what a model represents its target system? The accuracy question is the question of what makes a model accurately represent its target system? These two questions are constitutively related. Unless a model is taken to represent its target system, it does not make sense to ask to what extent it represents its target system accurately. As Nguyen and Frigg highlight: "The history of science provides us with a wealth of examples of inaccurate, but nevertheless representational, models. [Ptolemy's] model of the solar system represents the solar system even though it is inaccurate with respect to the orbit of the Earth around the Sun" (8). A model can represent its target system but with fluctuating degrees of accuracy (or, perhaps with no accuracy at all).

The standard view in the literature is that scientific models are representational models of their target systems in the world. Furthermore, part of the standard view is that models are (or will tend to be) inaccurate models when compared across the model-world relation. This can also be said of the examples of models introduced in sections 1.1 and 1.2. However, if we are working with the idea that these models are inaccurate models vis-à-vis their intended target systems, we can also say that they are not inaccurate for the same reasons. The small-scale model of the SS *Monterey* represents its target system, the actual SS *Monterey*. However, it represents its target system inaccurately because it omits some aspects of the SS *Monterey*. The model does not include the layouts of the dining hall on the actual ship, and so on. This modeling technique is referred to as *abstraction*. All the computational and mathematical models mentioned above are inaccurate representations of their intended target system for different reasons, even though they all make use of abstraction. Computational and mathematical models make use of *idealization* in addition to abstraction. I made a point of this when I introduced the McCulloch and Pitts model. The standard view of idealization is that idealizations are the deliberate introduction of distortions of reality into a model for explanatory gain (McMullin 1985; Thompson-Jones 2005; Weisberg 2007). Thus, idealizations deliberately misrepresent their target systems. Because idealizations misrepresent their target systems, they are defined as falsehoods.

Idealization in science has generated a lot of attention recently. A dominant view is that idealized models are akin to fictions, motivating a particular brand of fictionalism about scientific models (see, e.g., Frigg 2010;

Godfrey-Smith 2009; Toon 2023). Fictionalism about models gives rise to a host of important questions. For example, if scientific models are fictional, and if these models are ubiquitous in science, does idealization undermine the realist quest for truth? Further issues are: Are scientific models qua abstract models merely the product of the imagination—a kind of sophisticated make-believe (for discussion, see Frigg 2010)? Or, how can idealizations provide explanations of real-world phenomena if they are false with regard to those phenomena (see, e.g., Elgin 2017)? These sorts of questions have received most attention in the context of mathematical models qua abstract (nonconcrete) models. In this book, I treat computational and mathematical models equally. Insofar as both types of models make use of idealizations, which they do, there is no reason why all the questions raised about the status of mathematical models cannot be raised for computational models. As I mentioned, these techniques are both in use in computational models. This should come as no surprise. Life does not evolve as it does in the Game of Life.

Most computational and mathematical models, as I mentioned in the introduction, are abstract theoretical devices that exist (mainly) in the imagination of the scientific communities who make use of such models to understand the world. What the Schelling model, the Game of Life model, and the HH model all have in common is: what they state is true of the model, but not necessarily of the world. What do we learn from models such as Schelling's? What is the lesson of the Game of Life? It is neither that residential segregation unfolds according to a single preference function nor that life is a cellular automaton. Yablo suggests that one lesson of the Schelling model is that "racial segregation could result from dislike of isolation, as far as the statistical evidence goes" (2020, 143). The Game of Life suggests that were we to be isolated we would likely die—or, alternatively, overcrowding is counterproductive to survival.

1.5 The Map Problem: Scientific Realism vs. Antirealism

So far, we have considered questions such as: What is a scientific model? In the literature, three types of models are usually identified. These models are: concrete, computational, and mathematical. One can also ask: How does model building proceed? Building models has a set of signature steps. Start by taking some pattern of behavior you wish to explain—the target system. It might be residential segregation, action potentials, the origins of life, maze navigation, locomotion, and so on. This might initially involve formulating a mathematical description of a model, as we saw in the case of the HH model (Weisberg 2006). Then test the model for internal coherence. Finally, and if warranted, assess the relationship between the model and its target system.

The question of interest here is: What do scientific models allow us to say about the features of a target system? That is, how could an idealized model that is so divorced from reality provide epistemic insights? The McCulloch and Pitts model is not at all like actual biological neurons. So how could such a model facilitate explanation and understanding of the nervous system? The Schelling model is not like actual cities. So, again, how might such a model help us explain residential segregation? Similarly for the Game of Life model, the HH model, and so on. Or, to put it differently, if scientific models are akin to maps, which many take them to be, then what is the relation between models and the realities they are invoked to represent, given the relation between a map and the territory of which it is a map? I call this the *map problem*.

Addressing the map problem gives rise to the controversy between scientific realism and scientific antirealism. Scientific realism is, broadly speaking, the view that science aims at discovering the true nature of reality. The aim of science is not merely making sense of what we can observe. It is to explain the underlying nature of what we can observe (see, e.g., Godfrey-Smith 2003; Psillos 1999). By contrast, scientific antirealism is, generally speaking, the view that scientific progress is made primarily with regard to things we can observe and that such progress can be made without discovering truths about the unobservable parts of nature (see, e.g., Rowbottom 2019.

We can study models for their own sake. Any model that can be treated realistically can also be treated instrumentally, or antirealistically. This is especially the case when we consider certain kinds of models: *targetless models*. In this case, the only real concern is the study of the model itself. No claims are being made from the model system to the target system. This type of modeling is, following Weisberg (2013), similar to pure mathematical analysis (see also Nguyen and Frigg 2022 for discussion).

Targetless modeling is, therefore, a safe space for realists and antirealists alike. Or, realist and antirealist can agree or disagree about whether or not the mathematics is justified, or whether the initial assumptions of the model are appropriate, or whether there is sufficient internal coherence between the parts posited by a certain model, and thus like. The Game of Life is an example of this kind of modeling practice. Realists and antirealists can in principle agree that the computational structure underlying the Game of Life is Turing universal, even if a universal Turing machine is an idealization. They can agree that the algorithms specifying the state transitions in the Game of Life map onto the model system. That is, they can agree or disagree on how the model description maps onto the model system. Or, consider an example discussed by Churchland and Sejnowski (1990): the NETtalk system. NETtalk is a *connectionist* model. Such models are built to mirror (or approximate) the nervous system by focusing on connections and differential strengths of connections between processing units, where the processing units are intended to be like neurons. In the NETtalk model, there are three processing layers: an input layer, a hidden layer, and an output layer. The input layer receives letters as input, and the output layer produces speech sounds via a mediating layer of hidden processing units. NETtalk consists of a total of 309 processing units and 18,629 weighted connection strengths. The values of the weights are determined by a *back-propagation* learning algorithm. A back-propagation algorithm is an example of what is called a *steepest-descent algorithm*, where an "error signal"—the difference between an actual input and an anticipated input—is used to adjust the respective weighting of the hidden and output layer. According to Churchland and Sejnowski, if "taken literally as a model of functioning neurons, back-propagation is biologically implausible, inasmuch as error signals cannot literally be propagated back down the very same axon the signal came up" (1990, 366). Here, again, realists and antirealists can agree on the biological implausibility of the back-propagation algorithm. They can also agree or disagree about the internal coherence of a connectionist model that uses such algorithms, as long as we treat the NETtalk model as a targetless model.

The disagreement between scientific realists and scientific antirealists must, therefore, lie elsewhere. It arises when we turn our attention to two other kinds of modeling practices: modeling with a *generalized target* (generalized modeling) and modeling with a *specific target* (target-directed modeling). A generalized model might aim to understand something about sexual reproduction in mammals. Similarly, a computational model in neuroscience might have the general aim of understanding state transitions in nervous systems. A generalized model of sexual reproduction will, therefore, not say anything about sexual reproduction in wombats, for instance. Nor will a generalized model in neuroscience say anything specifically about the human nervous system. Target-directed models differ from generalized models in the following way: a model of sexual reproduction of the eastern gray kangaroo says something specifically about sexual reproduction in that species. A computational model targeting chemotaxis in single-celled organisms such as *Escherichia coli* (*E. coli*) says something specifically about the role of chemotaxis in adaptive behavior. In both kinds of modeling practices the map problem arises: What is the relationship between scientific models and the realities these models purport to represent?

The issue is aptly captured by Korzybski's (1933) expression "A map is not the territory it represents, but, if correct, it has a similar structure to the territory, which accounts for its usefulness" (58). Similarly, a scientific model is not the target system it ideally represents, at least indirectly. This looks to be the situation we are stuck with. As Korzybski goes on to note: "If the map could be ideally correct, it would include, in a reduced scale, the map of the map; the map of the map, of the map; and so on, endlessly" (58). Such an ideal map would be useless. This point is elegantly captured by Borges (1999), who wrote the following one-paragraph short story about the unrealistic burdens placed on the cartographers in a distant kingdom tasked with having to create an exact map of that kingdom. Borges's piece is entitled "On Exactitude in Science," and reads as follows:

> In that Empire, the Art of Cartography attained such Perfection that the map of a single Province occupied the entirety of a City, and the map of the Empire, the entirety of a Province. In time, those Unconscionable Maps no longer satisfied, and the Cartographers Guilds struck a map of the Empire whose size was that of the Empire, and which coincided point for point with it. The following Generations, who were not so fond of the Study of Cartography as their Forebears had been, saw that that vast Map was Useless, and not without some Pitilessness was it, that they delivered it up to the Inclemencies of Sun and Winters. In the Deserts of the West, still today, there are Tattered Ruins of that Map, inhabited by Animals and Beggars; in all the Land there is no other Relic of the Disciplines of Geography. (1999, 325)

The map problem is, therefore, here to stay. Here is how it relates to the discussion about scientific realism. Scientific realism is the view that scientific models say something true, or approximately true, about the nature of their target systems in the world, concerning observables and unobservables—even if models are not their target systems. Antirealists deny that scientific claims can be true, or approximately true, with respect to unobservables. Or, scientific antirealists stress that the aim of science is not truth. On this latter view, scientific models are nothing but instruments for the prediction and systematization of observables.

1.6 Conclusion

In this opening chapter, I have introduced some concepts, issues, and distinctions that will be of relevance throughout this book. Specifically, I mentioned that model-based science uses three different types of models: concrete, computational, and mathematical models. I then highlighted that the standard view of scientific models is that they are *representational* models of their intended target systems. However, even though models function—on the standard view—to represent systems in the world, they will often represent those systems inaccurately. Models will either omit features of target systems (abstraction) or they will function as deliberate distortions of target systems for explanatory gain of some kind (idealization). I ended by focusing on where in the discussion of scientific modeling the dispute between scientific realism and scientific antirealism arises. I located this dispute with the map problem: What is the relation between models and the realities they are invoked to represent, given that the map (or model) is not a perfect model of its territory (or target system)? In other words, the discussion about scientific realism gets going only when assertions are made from model descriptions to aspects of target systems in the world.

2 Idealization: The Standard View

Overview: In chapter 1, I present the view that scientific models are representational models, although either they leave out certain details of the target system or they are inaccurate (in some way), or both. Some models omit features of a target system. This modeling technique is called *abstraction.* All scientific models are instances of abstraction, for the simple reason that a model is not equal to its target system. Models simplify their target systems. Other models are inaccurate given the widespread use of idealization in scientific modeling. The standard view of idealization is that this involves the deliberate introduction of a distortion of reality into a model, usually for the purpose of explanatory gain (of some kind). Therefore, the standard view is committed to the idea that idealizations are a particular kind of misrepresentation. By implication, because idealizations are misrepresentations of target systems, they are falsehoods when compared across the model-world relation. The aim of this chapter is to introduce the standard view of idealization. Take a simple example. Each morning, I boil a kettle of water to make a cup of tea. My kettle contains a finite number of particles, and it has a finite volume. However, as Fletcher et al. (2019) note, it is often claimed that our best theories describe boiling and other phase transitions as occurring only in infinite systems. According to Kadanoff: "The existence of a phase transition requires an infinite system. No phase transitions occur in systems with a finite number of degrees of freedom" (2000, 238). Yet, phase transitions appear to occur everywhere in nature, such as melting snow (solid to liquid), freezing ice (liquid to solid), and evaporation from my hot cup of tea (liquid to gas). Brochini et al. report on phase transitions in the nervous system, where neuronal avalanches "occur whose distributions of size and duration are given by power laws, as observed in biological neural networks" (2016, 1). And yet, if Fletcher et al.

(2019) and Kadanoff (2000) are correct, we can understand phase transitions as occurring only in highly idealized models. My kettle does not have infinite volume, after all. Idealizations, therefore, are distortions of reality that enter "scientific theories, laws, models, or, more generally, representations" (Fletcher et al. 2019, 1657).

2.1 Introduction to Abstraction and Idealization

The aim of this section is to introduce abstraction and idealization in a little more detail. I will do so by considering the following example, from Thompson-Jones (2005). One afternoon, a cannon is put on an open plain and fired. A model is constructed to help us predict where the cannonball will land and ideally explain why it will land where it does. In building our model, we make several assumptions:

- The path of the cannonball is contained wholly within a perpendicular plane to the ground.
- A Cartesian graph is used to coordinate this plane such that the x-axis represents the ground, and the y-axis represents points vertically upward.
- The cannonball, when fired, moves so that its x-coordinate increases.
- The cannonball has some initial velocity of v when fired, and is, at that starting point, moving at an angle α to the x-axis.

Once the cannonball has been fired, we make additional assumptions: the cannonball only moves under the influence of gravity; and only gravity exerts a force vertically downward with a magnitude of $mg=9.8\text{m/s}^2$ throughout its motion.

According to Thompson-Jones:

> Given specific values of v and α, it is then a matter of simple integration to calculate the time of flight of the cannonball, the distance from the cannon at which it will land, the maximum height it will reach, and other features of its trajectory. And it is not much more difficult to show that, according to this model, the range of the cannon is maximized for a given v when $\alpha=45°$. (Thompson-Jones 2005, 182)

The model contains both abstraction and idealization. It abstracts by omitting certain features of the target system. As Thompson-Jones states:

> There are innumerable features of the modeled system which the model omits, without thereby misrepresenting or distorting the system: no mention is made of

> the composition of the cannonball, of its internal structure, of its color or temperature, of the composition of the ground. . . . The model is simply silent in all these respects. (Thompson-Jones 2005, 184)

The model idealizes in numerous ways. Specifically:

- It assumes that gravitational force has the same magnitude and direction at all points.
- It assumes that only the Earth exerts gravitational force.
- It assumes that the only force acting on the cannonball is gravitational.
- It assumes that the trajectory begins at the ground level (ignoring the height of the cannon). (Carillo and Knuuttila 2022, 53)

Moreover, it idealizes in the sense that its mathematical derivations are not part of the target phenomenon itself since the derivations are used to model the phenomenon, but are not part of the phenomenon itself. The idealizations are deviations from reality. This is the standard view of idealization. It takes idealizations to falsely represent parts of reality, or parts of target systems.

What is the relationship between abstraction and idealization? All idealized models use abstraction. Not all abstractions are idealizations. A small-scale model of the ship SS *Monterey* utilizes abstraction. Model parameters are selected, and tested, for similarity with the parameters that specify the SS *Monterey*. The HH model is an idealization and, therefore, an abstraction. We know this from chapter 1, where I said that the HH model of the action potential represents different types of ion currents, all of which I claimed to be explanatorily relevant for understanding the flow of those ions through the cell membrane. Yet, the HH model does not have ion channels. Ionic channels are proteins. The HH model is a set of equations (Brette 2013). One can identify parts of the equations that underwrite the HH model with properties of proteins. However, there are no real proteins in the HH model. There is something one can call proteins in the equations, but what we are referring to are not proteins in any real sense. Similarly, one can build a wooden plane and say that it has a helix. We can call it a helix. Yet, it is a "nice-looking piece of wood" (Brette 2013). The same goes for the HH model. We can call parts of the equations proteins, even though they are "merely" mathematical elements.

2.2 The Standard View of Idealization

I call the standard view of idealization the *falsehood view of idealization* (FvI). According to the FvI, idealization is understood as distortion. An idealization represents aspects of target systems differently than how they actually are. Specifically, because idealization represents aspects of target systems different than how they actually are, *misrepresentation* becomes central for the understanding of the nature and function of idealization. Moreover, because idealization is a form of misrepresentation, it is a *falsehood.* Misrepresentation and falsehood are the two fundamental components of how idealization is understood given the FvI. Here is a sample of researchers taking idealizations to be falsehoods (emphasis added to all quotes).

- Appiah: "Treating the model as an idealization means that we don't have to worry that it gets things wrong. For, again, an idealization is *just a kind of useful fiction.* And a fiction is something that *isn't true*, but that we treat for certain purposes as if it were" (2017, 73).
- Elgin: "When effective, models and idealizations are, I contend, felicitous *falsehoods*" (2017, 1).
- Elliott-Graves and Weisberg: "If a model is idealized relative to its target, then its representation of the structural features of the target may well be *distorted*" (2014, 181).
- Godfrey-Smith: "Treating things as having features they clearly do not have" (2009, 47).
- Lawler: "Science is replete with falsehoods that epistemically facilitates understanding by virtue of being the very *falsehoods* that they provide" (2021, 6859).
- Potochnik: "Idealizations are assumptions made without regard for whether they are true and often with full knowledge that they are false" (2017, ix).
- Shech: "Idealizations . . . are distortions or *falsities*" (2023, 1).
- Strevens: "A model of projectile trajectories . . . might be idealized . . . so *falsely* representing air resistance as entirely absent" (2019, 1715).
- Thompson-Jones: "My starting point, then, is the suggestion that we should take idealization to require the assertion of a *falsehood*" (2005, 175).

- Wimsatt: "All theories, even the best, make idealizations or other *false assumptions* that fail as correct descriptions of the world" (2007, 94).

Because idealizations are deemed to be falsehoods, those working with the FvI consider idealization to be a species of misrepresentation. Here is a small but representative sample from the literature on the misrepresentational part of the FvI (emphasis added in all quotations):

- Elgin: "Models do not, of course, only omit. If they did, there would be no reason to consider them *inaccurate*. They could be construed as schematic or partial representations of the phenomena they bear on. But models also *distend, distort, exaggerate*, and even introduce elements that answer to nothing in the target" (2017, 265).
- Elliott-Graves and Weisberg: "Such a model will contain unintentional *misrepresentations* that must be corrected. Many other idealizations, however, are legitimate" (2014, 179).
- Kaplan: "By 'idealization' I mean the deliberate introduction of simplifications and *distortions* into models for the purposes of making them more computationally tractable or to better model the dominant features of the phenomenon of interest" (2011, 348).
- Potochnik: "My view sanctions idealization even of central causal influences, on a permanent basis, and without taking any steps to hold in check the resulting *misrepresentation*" (2017, 59).
- Rice: "Our best scientific models and theories are pervasive *misrepresentations* of their real-world target systems . . . in agreement with Elgin, I will grant that the representations produced by scientific inquiry are not true and do not purport to be true" (2021, 4098).
- Strevens: "Because asymptotic idealization is characterized with reference to the process by which an idealized model is created—that is, some sort of limiting process—it will be illuminating to offer a procedural characterization of simple idealization too, identifying it not so much by the structure of the end product, a certain deliberately *misrepresenting model*, but by the way in which the structure is derived" (2019, 1715).
- Thompson-Jones: "The proposals I have just put forward provide only partial guides, for each lay down only a single necessary condition on correct application of the relevant term: *misrepresentation for "idealization,"* and omission without misrepresentation for "abstraction" (2005, 185).

- Weisberg: "Often a theorist will specifically introduce these *misrepresentations* to compensate for some other misrepresentation of the target, in which case they would be included in the feature set. In any case, intentional distortions can automatically be included among those features that are not shared between model and target" (2018, 256).

The example discussed in section 2.1 is an example of the FvI. Indeed, in the quote from Thompson-Jones (2005), there is the explicit identification of idealization with misrepresentation and thus of idealization being false of the world. In the next section, I further introduce the FvI and explore some reasons to idealize.

2.2.1 The FvI and Reasons to Idealize

Examples of idealization are not hard to find. Standard examples include frictionless planes, infinite number of particles, and continuously increasing and decreasing populations. Less obvious candidates are common in computational neuroscience. For example, continuous time is usually approximated in computational models of neural systems with a series of discrete time steps, enabling tractable simulations on digital computers. Synaptic integration is modeled by a simple linear weighted sum of a unit's input in neural networks. Furthermore, as Kaplan notes, "cable theory idealized dendrites and axons as perfectly cylindrical, passive conductors" (2011, 348).

A quite natural question to ask is: Why do scientists make use of assumptions about target systems that we know to be obviously false? Two different accounts of idealization have emerged in the literature as a result of answering this question. One is the nonpermanent view (also called the dispensability view, deficiency view, Galilean idealization, or heuristic idealization). The other is the permanent view (also called the indispensability view or essentialist view of idealization). The former holds, roughly, that idealization is temporary or nonpermanent. The latter asserts that because of the epistemic benefits provided by idealizations, they have a more permanent status in scientific modeling (see figure 2.1).

Galilean idealization is an example of the nonpermanent view of idealization. This view regards idealization as an instrument for gaining computational traction on complex target systems. Yet, as science progresses and technology develops, a purely de-idealized—that is, veridical—explanation of target systems will be possible. According to Weisberg: "The practice

	Issues with the world studied	*Issues with scientific practice and researchers*
Permanent idealization	- Computationally intractable problems - Generalizability - Sharpen focus on select features	- Cognitive limits - Pedagogical value
Non-permanent idealization	- Technological limitations - Complex target systems	- Current limits in computational power - Common-practices in a field of research

Figure 2.1
Reasons for idealization (modified from Potochnik 2017, 48).

is . . . nonpermanent. Galilean idealization takes place with the expectation of future de-idealization and more accurate representation" (2007, 642). The basic idea that a more accurate representation will be gained once idealizations are removed from scientific models clearly indicates that Galilean idealization is wedded to the idea that idealizations are a kind of misrepresentation of the world.

The causal difference-making view of idealization formulated by Strevens is an example of the permanent view. It treats the function of idealization as highlighting which salient causal factors are explanatorily irrelevant by flagging their "failure to make a [causal] difference" (2008, 318). According to Strevens, what makes for a good idealizing explanation is that it conveys the most essential information—information about causal difference makers—and "is always better than its veridical counterpart" (2008, 318). If Strevens is correct, then the aim of de-idealization will likely decrease the model's explanatory value. The causal difference-making view of idealization is an example of the FvI, since the idealized model delivers scientists with information about which factors do not play a causal difference-making role, setting them aside by misrepresenting them—for example, in the ideal gas law molecular collisions are set to the value of zero, marking this parameter as a causal non-difference-maker. Strevens is explicit about identifying idealization as a kind of misrepresentation, when he states: "I define idealization as any misrepresentation of the causal process that improves the explanatory power of a model by comparison to its veridical counterpart" (300).

The holistic view of idealization (Batterman and Rice 2014; Rice 2018, 2019) marks a different permanency view of idealization. Unlike Strevens's (2008) account of idealization, which presupposes that scientific models can be cleanly decomposed into the contributions made by their accurate and inaccurate parts, respectively, the holistic view denies this decomposability strategy (see Rice 2019 for details). The holistic view is motivated by the idea that different kinds of idealizing assumptions are required when working with computational, mathematical, and statistical techniques in constructing scientific models, making idealization an essential part of model-based science. I agree with this basic insight, and discuss the implications of this in chapter 9, when I address the use of idealization in formulating a grand unified theory of the brain. However, the holistic view is problematic because it conforms to the FvI. As Rice notes: it is "the pervasive *misrepresentation* of the features, processes and entities of the model's target system" that enables scientists to "extract the desired explanatory information that would otherwise be inaccessible" (2018, 2809; emphasis added).

Thus, regardless of whether one considers permanent or nonpermanent accounts of idealization, the standard view is that given their role in scientific modeling, idealizations function as means by which to represent target systems. On the representational view, therefore, idealizations will inevitably turn out to misrepresent their target systems. Because they misrepresent their target systems, they are false when considered across the model-world relation. Thompson-Jones puts this aptly when he states: "We habitually lie about some aspects of the systems in question, and entirely neglect to mention others" (2005, 174). The FvI is, therefore, the view that scientific models violate the whole truth about target systems in virtue of misrepresenting them.

The standard view of idealization treats idealizations not merely as false, but as useful falsehoods. The specific issue that I take issue with in the next chapter is that we "habitually lie about some aspects of the systems in question." In my view, idealizations are not lies. It is not the case that idealizations are known to be false by those using them but pretended to be true to deceive an audience (Mäki 2020). The best way of capturing what I find problematic about the FvI is the idea that idealizations are not simply false, but deliberate falsehoods, in the sense that those using idealizations are wholly aware of their falsity. The very idea that idealizations are deliberate falsehoods rests on the view that idealizations deliberately misrepresent aspects of reality. However, for idealizations to come out as misrepresentations of reality, their content must be taken literally. It is this literalist assumption that pervades discussions of idealization that I find problematic.

2.3. Conclusion

In this chapter, I have described how abstraction and idealization are usually understood in the literature on scientific modeling. Abstraction is the practice of building simplified models of target systems by leaving aside or omitting features of the target system. I paid special attention to formulating what I take to be the standard view of idealization, which I called the "falsehood view of idealization" (FvI). The FvI presupposes that a central feature of idealizations is that they misrepresent aspects of their intended target systems. Because they misrepresent, idealizations are falsehoods. Idealizations are useful and legitimate; yet, they are falsehoods, nevertheless.

3 Idealization: The Nonliteral View

Overview: The aim of this chapter is to argue that the FvI is problematic. The FvI is premised on the assumption that idealization is a way of deliberately introducing distortions of reality into a scientific model. For idealizations to come out as falsehoods, scientific discourse about idealization must be taken literally. When viewed through the lens of literalism, idealizations cannot be anything but falsehoods—even if they are epistemically useful.

In this chapter, I argue that the function of idealization is not to represent aspects of target systems; nor is it the function of idealization to serve as a useful legitimate falsehood. I argue that idealizations are *not* falsehoods provided we understand idealizations nonliterally. A nonliteral treatment of idealization has the implication that idealization does not serve any representational function across the model-world relation. Idealizations are, I argue, without classical truth-values. Or, differently put, idealizations are nothing but scientific instruments or tools.

3.1 Introduction

Idealizations are widespread and pervasive in science. It is common to represent or define idealizations as deliberate distortions or falsities, which scientists enter into theories, laws, models, or scientific representations more generally (from chapter 2). As we know, standard examples include: the ideal gas law, frictionless planes and infinite biological populations. Other cases include the view that the brain is a computer, that cognition is a control system like Watt's centrifugal governor, or that mental representations have a language-like structure captured in a language of thought.

It is equally widespread and pervasive to view idealizations as useful falsehoods. In chapter 2, I called this the "falsehood view of idealization." The FvI is based on the view that idealization is a falsehood, since the FvI is committed to the idea that misrepresentation is central to a proper understanding of idealization. Rejecting the claim that idealizations are falsehoods implies that they are not misrepresentations of target systems, or of aspects of target systems. Conversely, rejecting the claim that idealizations are not misrepresentations implies that idealizations are not falsehoods.

In this chapter, I argue that idealizations are useful and legitimate. However, idealizations are not falsehoods and, therefore, are not misrepresentations of the world. The FvI rests on the view that scientific discourse must be taken literally, that is, scientific claims must be understood as truth-conditioned descriptions of some specific target domain (literalism). This means that for idealizations to come out as falsehoods, literalism must be accepted about scientific discourse involving idealizations. Accepting literalism about scientific discourse is to accept a version of what is called *semantic realism*. There is no single agreement on how to formulate semantic realism (see Miller 2003 for discussion). Here I will restrict myself to the basic idea that scientific discourse is truth-conditioned—that is, scientific claims must be either true or false. This is how semantic realism appears in most formulations of scientific realism; namely, that we take scientific discourse at face value—as literal descriptions of their intended subject matter (Chakravartty 2017; Psillos 1999; Rowbottom 2022).

Treating idealizations nonliterally undermines the FvI. It also forces one to rethink semantic realism as it appears in formulations of scientific realism. I address only the former in this chapter. I address how to reformulate semantic realism in a defense of scientific realism in chapter 4. Rejecting the FvI renders idealizations useful and legitimate parts of scientific practice, without idealizations being falsehoods and inaccurate representations of their intended target domain. The precise reason is that when treated nonliterally, the function of idealizations can be shown not to consist in representing parts of the target systems that scientific models are invoked to explain.

In the debate about scientific representation a standard question is: What makes a model represent its target system? A slightly different question is: What makes a model an accurate representation of its intended target system? Following Nguyen and Frigg (2022), I labeled these questions the "semantic question" and the "accuracy question," respectively, in chapter 1. For both questions, if the FvI is assumed, idealizations will turn out as falsehoods. Evidence for this is provided in chapter 2. Understanding idealizations nonliterally shifts this focus on the falsehood of idealization toward a different discussion entirely. The nature and function of idealization in scientific models, I submit, are not to represent anything about real-world target systems. Rather, the functions of idealizations simply are

- to flag that certain variables are causally irrelevant to explanation;
- to sharpen our focus on specific causal factors, rather than others; and
- to facilitate computational tractability.

These three functions of idealization justify their usefulness in science. However, to describe idealizations as useful legitimate falsehoods rests on the unnecessary assumption that all scientific claims must be understood as truth-conditioned descriptions of some specific target domain. In this chapter, I reject literalism about some scientific discourse, and defend the view that discourse about idealizations should be understood nonliterally.

I proceed as follows. In section 3.2, I give three examples of the FvI. For each example, I highlight the specific function the idealizations are intended to serve. In section 3.3, I move on to develop the nonliteral view of idealization. I argue that for each of the three cases discussed, a nonliteral interpretation of the relevant idealization is theoretically preferable to a literal interpretation. In section 3.4, I consider other nonliteral proposals in the literature, especially the work of Frigg and Nguyen (2021) and Nguyen and Frigg (2022). There is an undeniable similarity between the nonliteral view of idealization and the DEKI (denotation, exemplification, keying-up, and imputation) model developed by these authors. Like Frigg and Nguyen (2021), I deny that idealizations *denote*—that is, idealizations do not represent their target systems as actually having idealized properties such as infinity, no air resistance, and so on. I depart from Frigg and Nguyen (2021) in the following ways. (1) Frigg and Nguyen (2021) develop their nonliteral view of some scientific discourse—when it involves idealization—in response to Elgin (2017). I identify that literalism about idealization plagues the entire field involved in the discussion about idealization (chapter 2). (2) Unlike Frigg and Nguyen (2021), I put the nonliteral view of idealization to use in a defense of scientific realism. (3) I apply the nonliteral view of idealization to draw several implications for scientific modeling in the cognitive sciences. So in section 3.4, I focus especially on spelling out the DEKI account. I conclude in section 3.5.

3.2 The FvI and Three Case Studies

In this section, I briefly describe three case studies that have been used to support the FvI. For each case study, I highlight which precise function the invoked idealization has. Some of the cases involve more than one function of idealization. However, for simplicity, I treat each of the three case studies as exemplifying one particular function of idealization.

3.2.1 The Ideal Gas Model

According to Strevens (2008), almost any formal construct can serve as a model. A set of equations such as the HH model, a computer simulation of weather patterns, a mechanistic representation of an action potential, or a working model of the solar system (Strevens 2008, 16). A much-discussed scientific model in the literature is the ideal gas law (or model). The ideal gas law is represented as follows: $PV = nRT$. P is pressure measured in atmosphere; V is volume measured in liters; n is moles of gas present (a mole is a standard unit in chemistry used to measure large quantities of very small entities such as molecules); R is a constant of the units of moles; and T is temperature measured in degrees Kelvin. The ideal gas law enables one to relate the macroscopic properties of gases by considering the pressure, volume and temperature of a hypothetical ideal gas. In this case, an ideal gas is a gas in which the molecules (1) do not repel or attract one another and (2) do not take up any space (i.e., they have no volume). In her discussion of idealization, Elgin states:

> The ideal gas law, $pV = nRT$, represents the relation between pressure, volume, and temperature in a gas comprised of dimensionless, spherical molecules that exhibit no mutual attraction. *There is no such gas; indeed, if our fundamental theories are even nearly right, there could be no such gas.* All material objects have spatial dimensions. Being subject to gravity, all attract one another. No molecules are spherical. Nonetheless, the ideal gas model is integral to thermodynamics. (Elgin 2017, 15; emphasis added)

No natural gas is an ideal gas. According to the FvI, the ideal gas law deliberately distorts how it represents natural gases. The idealizations serve a particular function: they *downplay* the causal relevance of intermolecular interactions in explaining the properties of natural gases. Strevens agrees:

> My proposal: the assumption that there are no intermolecular collisions is intended to communicate to the reader that collisions have no net effect on the relationship between volume and pressure. One way to say that collisions have no net effect is to say that it is as if there are no collisions; this is how the idealizing explanation is to be understood. (Strevens 2008, 316)

According to the FvI, the idealizations in the ideal gas law are deliberate falsehoods; yet these falsehoods or distortions of reality are epistemically useful in the sense that the ideal gas law allows for a good enough approximation of natural gases.

3.2.2 The Hardy-Weinberg Model

Idealizations introduce features into a model that distort or misrepresent the target system. In the ideal gas law, the modeling assumptions that molecules (1) do not repel or attract one another and (2) do not take up any space (i.e., they have no volume) are examples of such distortions. Their function is to flag that (1) and (2) are not directly causally relevant for explaining the behavior of actual gases. Hence, (1) and (2) are idealized away. Here I want to consider a different model, the Hardy-Weinberg model. It too is an idealized model. However, its function is not to flag the explanatory irrelevance of certain aspects of the world. Rather, the idealizations serve to sharpen our focus on particular causal factors.

The Hardy-Weinberg model describes the unchanging frequency of alleles and genotypes in a stable, idealized biological population. An allele is one of two (or more) versions of DNA sequence at any given genomic location. Let us consider how the model is presented by Elgin:

> [The] Hardy-Weinberg model . . . represents the redistribution of alleles in the absence of evolutionary pressures. It tracks alleles in an infinite population of organisms that mate randomly and are not subject to mutation or migration. That the population is infinite counteracts the effects of genetic drift; that mating is random ensures that neither physical proximity nor genes that give rise to attractive phenotypes have a preferential advantage; that neither mutation nor migration takes place ensures that no novelties are introduced into the gene pool. The model is no approximation. Populations are not nearly infinite (what ever that might mean). Mating is not nearly random. However indiscriminate actual mating behavior is, physical proximity is required. In the long run, mating only with nearby partners promotes genetic drift. Natural selection and genetic drift are ubiquitous. Migration is widespread. Mutation and random fluctuations are, in real life, unavoidable. Still, to understand the effects of evolution, it is useful to consider what would happen in its absence. By devising and employing an *epistemically felicitous falsehood*, biologists find out. (Elgin 2017, 29; emphasis added)

In this model the function of idealization is not to highlight that migration, mutation, population limits, and so on, are causally irrelevant to understanding the distribution of alleles downstream. In the Hardy-Weinberg model, the idealizations serve the function of isolating certain causal mechanisms or dependency relations. As Mäki notes: "Many models in natural and social sciences are minimal models in this sense: they help focus on the structure and operation of one mechanism at a time, in isolation from all other mechanisms and conditions that in actual circumstances may also shape the behavior of the phenomena under investigation" (2020, 219).

3.2.3 An Idealized Model of the Trajectory of a Baseball

Idealizations either flag that certain variables are causally irrelevant to explanation, or they help isolate specific causal factors. A third function of idealization is that they facilitate computational tractability. According to Piccinini:

> Mechanistic explanations are often presented in the form of (interpreted) mathematical or computational models—e.g., the Hodgkin–Huxley model of the action potential. Typically, such models become intractable or difficult to manipulate if they include too many details about their target systems. Instead, mathematical and computational models are typically constructed not only by abstracting away from many details of the target system, but also by replacing those details with simplifications and idealizations that distort or misrepresent the target system. (Piccinini 2020, 171)

Idealizations, therefore, often serve to solve analytically intractable problems. Here is another example, from a physics textbook that illustrates how physicists conceive of idealization in moving from a target system to the generation of an idealized model in accounting for the behavior of the target system. As Young and Freedman state:

> *In physics a model is a simplified version of a physical system that would be too complicated to study in detail.* . . . Suppose we want to analyze the motion of a baseball thrown through the air. How complicated is the problem? The ball is neither perfectly spherical nor perfectly rigid; it has raised seams, and it spins as it moves through the air. Wind and air resistance influence the motion, the earth rotates beneath it, the ball's weight varies a little as its distance from the earth changes, and so on. *If we try to include all these things, the analysis gets hopelessly complicated.* Instead, we invent a simplified version of the problem. We neglect the size and shape of the ball by representing it as a point object, or particle. We neglect air resistance by making the ball move in a vacuum, we forget about the earth's

> rotation, and we make the weight a constant. *Now we have a problem that is simple enough to deal with.* (Young and Freedman 2007, 3; emphasis added)

In the idealized model, the baseball is represented as a point object, that is, as dimensionless and without mass. There are no point objects in the natural world. They are imaginary, theoretical constructs. Moreover, it is assumed that there is zero air resistance. This is of course false of the actual baseball. From the perspective of the FvI, the idealized model misrepresents its target system. However, doing so facilitates performing the relevant computations.

3.3 The Nonliteral View of Idealization

The aim of this section is to develop and defend the following claims: (1) idealizations are useful and legitimate, and (2) idealizations are not falsehoods.

3.3.1 Idealizations Are Useful and Legitimate

Everyone in the literature agrees that many idealizations are useful and legitimate tools in scientific modeling. Moreover, whether or not one is working with the FvI, everyone pretty much agrees on the particular functions of idealization in scientific modeling. In other words, there is consensus about the functions of idealization. The functions of idealizations are, to repeat:

- to flag that certain variables are causally irrelevant to explanation;
- to sharpen our focus on specific causal factors, rather than others; and
- to facilitate computational tractability.

In addition, one might also highlight that idealizations enable generalization, are required because of issues to do with cognitive limitations and that idealizations have certain pedagogical values (see, e.g., Potochnik 2017). The nonliteral view of idealization that I aim to develop and defend in this section subscribes to all these functions of idealization in scientific modeling.

3.3.2 Idealizations Are Not Falsehoods

A natural way to read the FvI is that this view of idealization treats idealized posits as factual propositions. As factual propositions, idealized posits are not only false—they are deliberate falsehoods and researchers using idealizations are aware of their falsehood. As Mäki puts it:

> Their falsehood is obvious, and it is accepted for a purpose. Idealizations are strategic falsehoods, deliberately employed by modelers to achieve their goals. This means idealizations are *not errors*, mistakenly believed to be true. They are *not lies* either (contra Cartwright 1983), known to be false by their utterers but presented as true to the utterers' audiences to deceive them. (Mäki 2020, 218)

According to the nonliteral view of idealization, it is a mistake to treat idealizations as deliberate falsehoods—as false assertions about the world. This view of idealization is to a reasonable degree shared by Mäki (2020). Mäki asks us to consider the assumption of a closed economy, relying on idealizations such as a country's exports and imports being equal to zero:

> This composite idealization can be used for making several other claims. . . . *Negligibility*: That the country's exports and imports>0 is negligible given the purpose for which a closed-economy model is being used. *Applicability*: The model applies to economies that are closed or whose deviation from full closure is negligible. *Tractability*: Assuming a closed economy enhances the mathematical or theoretical tractability of the model. *Early-step*: The closed economy idealization is part of an early version of the model that will be relaxed in later versions. None of these make the (false) claim that an economy is closed. These are alternative altered versions of a straight idealization that assumes some extreme perfection. These altered versions can be used for explicating the functions of idealizations and thereby for alleviating puzzlements around them. Note that a straightforward idealization, *if interpreted literally and considered a truth-apt claim, is known to be false*, while these altered versions have a chance of being true. (Mäki 2020, 220; emphasis added)

The nonliteral view rejects the position that idealizations should be interpreted literally. Thus, it rejects the view that idealization can be considered as any kind of *truth-apt claim*. The implication of this is that not all scientific discourse should be taken at face-value—that is, as either true or false—when assessed across the model-world relation. One can say that the use of an idealization is true in the model. It is true that the ideal gas model represents the ideal gas as involving zero molecular interactions. Yet, it is not necessary to also say that it represents actual gasses as having this feature.

Consider an example from Maddy (1992). She notes that the pages in any freshman physics textbook are littered with uses of mathematics that are explicitly understood not to be literally true. Maddy presents an example involving the analysis of water waves, "by assuming the water to be infinitely deep," and highlights that this "merely useful mathematics is still indispensable; without these . . . assumptions, the theory becomes unworkable" (281). Curiously, there are parts of the theory that provide a finite depth for oceans—namely, standard sonar techniques. Taken together, we have it that oceans can be modeled as both infinitely deep and not infinitely deep. Yet, we do not think that scientists subscribe to a set of contradictory beliefs: that oceans are both infinitely deep and not infinitely deep. So the idealization must serve a different function than representation. Or, the function of idealization is not to make truth-apt claims about reality. The idealized conditions in this model serve a tractability function—that is, assuming that oceans are infinitely deep enhances the mathematical or theoretical tractability and, therefore, precision of oceanic models.

Consider another example. It is an example from Walton (1990), and used by Toon (2023) in a discussion about mental fictionalism:

> Where in Italy is the town of Crotone?, I ask. You explain that it is on the arch of the Italian boot. "See the thundercloud over there—the big angry face near the horizon," you say; "it is headed this way." Plumbers and electricians distinguish between "male" and "female" plumbing and electrical connections. We speak of the saddle of a mountain and the shoulder of a highway. (Toon 2023, 16–17)

If someone asks you where the town of Crotone is, and you say "it's on the arch of the Italian boot," the person can either treat your description literally or the person can treat your description nonliterally. If taken literally, the description is false. Italy does not have a boot, after all. If taken nonliterally, the description is not taken as representing Italy as having a boot. Nonliterally, the description has a function: it facilitates navigation. However, it does not represent the country of Italy as literally having a boot.

I want to say the same for idealizations in scientific modeling. If taken literally, the ideal gas law represents natural gases as having zero intermolecular interaction. This is false. If taken literally, the Hardy-Weinberg represents that allele distribution is not impacted by migration and mutation. This is false. If taken literally, the idealized model of the trajectory and speed of a baseball would also be false. Similarly, if taken literally, the description that Crotone is on the arch of the Italian boot would be false.

One might object to the notion that physicists think that the idealized model represents a real ball (an actual target system). Here we need to be careful. The idealized model is invoked to explain the flight path of an actual ball. This much is true. Technically, the idealized model exemplifies the flight and trajectory of a ball as having certain features—lacking air resistance, being a point object, and so on. However, the idealized model fails to denote. Denotation is at the core of representation (Goodman 1976; see also Elgin 2017; Frigg and Nguyen 2021. Denotation is, paradigmatically, the extension of proper names, or all the things a word refers to. For example, one can use the name "Jonas Vingegaard" to denote a particular individual, namely Jonas Vingegaard. This is possible because there is an actual person that exists named Jonas Vingegaard—the winner of the 2023 Tour de France. In the literature, it is widely agreed that denotation can include such diverse elements as symbols, maps, charts, models, diagrams, and so forth. However, as Nguyen and Frigg highlight: "Denotation is a dyadic relation that obtains between certain symbols and certain objects. The relation holds *only if both relata exist*, implying that something that does not exist cannot be denoted" (2022, 54; emphasis added). Point objects do not literally exist in the world, even if they exemplify certain elements of an object that can be used to reason about the world. This is the case for all the idealizations we have looked at so far. Theoretical posits such as "infinite populations," "point objects," "zero molecular interactions," and so on, all fail to denote because they do not stand in a dyadic relation to anything in reality.

There is nothing unusual about this way of thinking about idealizations. Indeed, Elgin (2017) anticipates this kind of response, even if she adopts the FvI. Following Goodman (1976), Elgin (2017) distinguishes between "representations *of p*" and "*p*-representations." On this account, if *y* is a representation *of p*, then *p* exists, and *y* represents *p*. If a model represents electrons as having mass and charge, it does so in virtue of mass and charge being real properties of electrons. Now, *y* can represent *p* without being a representation *of p*: *y* may be a *p*-representation, where there is no corresponding feature of the world that is *p*. For example, a model may represent electrons as having self-rotation (i.e., spin) even if electrons do not actually have self-rotation. This would then be a way of idealizing an electron. If the distinction between "representations *of p*" and "*p*-representations" is correct, then representing an electron as having spin would be a *p*-representation. It follows that idealizations are not representations *of p*; rather, idealizations are *p*-representations. As Elgin states: "Thus, there are griffin-pictures, even though there are no griffins to depict. There is an ideal-gas-description even though there is no ideal gas to describe" (2010, 3). In short, we can say of maps, laws, models, diagrams, and so forth, that they exemplify certain features of a model even if they do not denote any aspects of a target system. This means that one can exemplify the town of Crotone as being on the arch of the Italian boot, or one can exemplify biological populations as being infinite, or one can exemplify gases as exhibiting zero intermolecular interactions, even though these features do not denote anything in reality.

This bears on the discussion in the following way. If idealizations function as *p*-representations, then one can give idealized posits as literal treatment when considering the scientific model that the relevant idealized posit is part of. When working with the ideal gas model it would be incorrect to say that the model represents collisions between gas molecules. Hence, *p*-representations can be true or false only when understood in the absence of making claims about reality. However, it would be false to conclude that the ideal gas model represents actual gases as having no intermolecular collisions. The reason for this is that the idealized posits fail to denote.

The problem with the FvI is that when idealization is discussed in the literature, idealization is presented as "felicitous falsehoods," "as strictly speaking false," "as entirely absent," or as failing "as correct descriptions of the world." Yet, this is to describe the representational function of idealization not from within the model itself but across the *model-world relation*. In other words, the FvI blurs the distinction between "representation *of p*" and "*p*-representation." This is also true of Elgin (2017), since for an idealization to be a useful falsehood, its status as a falsehood arises only if the idealization is taken to function as a "representation *of p*." The upshot of this is: if treated nonliterally, idealizations do not come out as falsehoods. That is, if *y* is a representation *of p*, then *p* exists, and *y* represents *p*. However, idealizations are not representations *of p*, because *p* does not exist. Infinite populations, ideal gases, frictionless planes, point objects, universal Turing machines, and so on, do not exist in reality. Their "existence" is limited to being parts of scientific models.

Let us consider a possible objection to the claim that idealizations do not serve any representational function across the model-world relation. The objection is that this view of idealization—as not representing parts of reality—is at least prima facie implausible. To appreciate the issue, let us consider the official map of the Bay Area Rapid Transit (BART) system in the San Francisco Bay Area (figure 3.1).

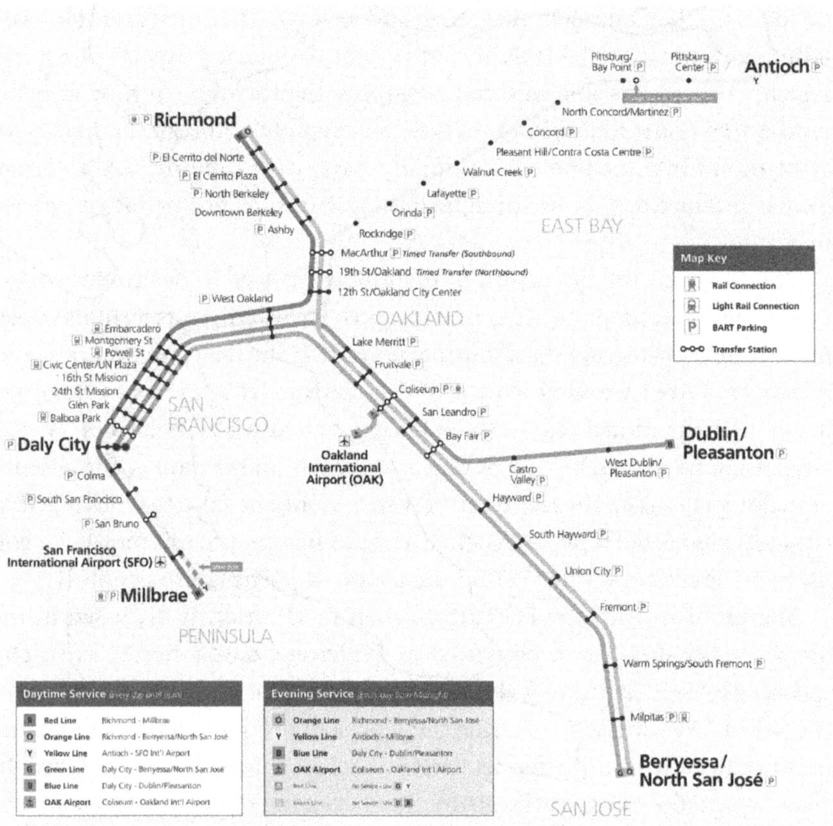

Figure 3.1
The official map of the Bay Area Rapid Transit (BART) system in the San Francisco Bay Area. Each rail line is represented by a different color and the stops on the map are equidistant and in a straight line. Downloaded from https://www.bart.gov/schedules/developers/maps.

Clearly, or so the objection goes, the idealizations in figure 3.1 serve to represent the realities of the BART system in the San Francisco Bay Area. Hence, the claim that idealizations do not represent aspects of reality is false. The conclusion is intuitive. Yet it does not follow. If we read the map literally, then it is clearly false. The map correctly represents South Hayward as coming after Hayward when traveling on the Richmond to Berryessa/North San José line. However, it is not the function of the map to claim that each stop is equidistant. That is, we would be wrong to conclude that all the stops in the BART system are equidistant by consulting the map. The idealized parts of the map, therefore, should not be taken to have this representational function. The idealizations of the map serve to organize our understanding of the world, that is, which station comes before or after another station depending on our direction of travel. The idealizations in the map simplify the structure of the map and add pedagogical value—they do not claim to represent all the stops as equidistant. Thus, the idea that idealizations do not represent aspects of the world is not implausible.

We can illustrate this further by considering a sentence *S* of the form "*p* and p_1 . . . and p_n and *J*" (Rowbottom 2022, 884). *S* is the proposition that "electrons have mass and charge and spin." I take *S* to be a possible placeholder for any scientific model. For example, we could write *S* as follows: "A cannon is placed on an open field and has a certain mass and only gravity exerts a force vertically downward with a magnitude of $mg = 9.8\ m/s^2$ throughout its motion." *p* and p_1 . . . and p_n can be given a literal treatment, in the sense that it is true that electrons have mass and charge. *p* and p_1 . . . and p_n all denote. Similarly, it is true that the cannonball is fired, is in motion, and so on. *J* is the claim that "electrons have spin" or that "only gravity exerts a force vertically downward with a magnitude of $mg = 9.8 m/s^2$ throughout its [the cannonball's] motion." *J* is an idealization. According to the FvI, *J* is false. Or, *J* is false because "intrinsic angular momentum" misrepresents an electron by ascribing spin to it (Rowbottom 2022)—when, in fact, it does not have spin.

There is room for discussion here, since you might find that the example provided by Rowbottom (2022) is a bit of a caricature. It is true that many first-year introductions to physics sometimes explain spin through an analogy with a rotating ball. Yet, it is also true that such introductions often caution the reader not to take this analogy literally. Indeed, once one starts to consult more advanced textbooks, the picture that is generally accepted is that spin is an intrinsic property of matter like mass, and that spin cannot be reduced to any other properties (like rotation). Whether or not Rowbottom's example is correct need not distract us. The point is: insofar *J* functions as a genuine idealization—e.g., that biological populations are infinite, or that natural gases exhibit no intermolecular interactions, or that edge detection in early vision involves computing the Laplacian of a Gaussian distribution—*J* will come out as false if interpreted literally. Hence, on the FvI, because *J* is false, *S* is inaccurate. Or, differently put, the falsity of

J implies that *S*—when understood as a conjunction of claims about the world—is false.

This is a counterintuitive outcome, since we would want to say that *S* is at least partially true, or approximately true given *p* and $p_1 \dots$ and p_n. One way to resist the conclusion that *S* is false is to give up on literalism about scientific discourse. If we give *J* a nonliteral treatment, then we can say of *J* that *J* is neither true nor false. If *J* does not serve any representational function, then *J* is not true about the world; nor is *J* false with respect to the world. Hence, *J* is neither true nor false when treated nonliterally from a model to the world.

The nonliteral view of idealization has several benefits that are not shared with the FvI. First, it allows for a view of idealization as both useful and legitimate, without having to justify how obvious falsehoods can have epistemic value. For example, Rowbottom notes that "the idea of spin (qua self-rotation) proved useful in making predictive progress, and especially in working out the proper (qua predictively successful) values for the fourth quantum number" (2022, 886). So, idealizations provide epistemic benefits, facilitating scientific progress. Unlike the FvI, it is not in virtue of their status as falsehoods that idealizations have the benefits that they do. Idealizations are beneficial only because of their specific functions in scientific modeling: (1) flagging that certain variables are causally irrelevant to explanation; (2) sharpening our focus on specific causal factors, rather than others; and (3) facilitating computational tractability. Second, the nonliteral view of idealization provides one way of understanding how scientific models can be said to be at least partially, and hence approximately, true of their target systems despite the rampant use of idealization in model-based science. Levy (2015) explains the notion of "partial truth" with reference to the ideal gas law. He states:

> It explains Boyle's law, namely: a gas's volume is inversely proportional to pressure. As the name suggests, the ideal gas model makes a number of idealizations—for instance, it assumes that gas particles do not collide with each other. But these assumptions can be shown not to affect the relationship between pressure and volume. That is to say: the question of inter-particle collisions turns out to be irrelevant to explaining Boyle's law. On the partial truth account, one evaluates the truth of the parts comprising a model, in terms of what the ideal gas model says of real gases. . . . When we do so, we find that it is partly true, i.e. with respect to the role of the motions of the particles. (Levy 2015, 793)

The nonliteral view of idealization implies that the ideal gas law is partially true, since it gets certain things correct about the behavior of real gases. Part of the ideal gas law is composed of nonidealizing, corresponding claims. However, it is also partly composed of idealizing, noncorresponding claims, where "noncorresponding" should be read as failing to denote and therefore failing to function as a representation of *p*.

3.4 The DEKI Account

The nonliteral view of idealization shares several features with the DEKI account of scientific representation developed by Frigg and Nguyen (2021) and Nguyen and Frigg (2022). DEKI—*denotation, exemplification, keying-up*, and *imputation*—is presented as follows (Nguyen and Frigg 2022, 62).

DEKI: Let $M=(X, I)$ be a model, where X is an object and I is an interpretation.

M represents T if and only if the following four conditions are satisfied:

1) M denotes T (and in some cases parts of M denote parts of T).
2) M exemplifies Z-features $Z=\{Z_1, \ldots, Z_m\}$.
3) M comes with a key K which associates the features in Z with another set $Y=\{Y_1, \ldots, Y_m\}$ of features: $K(\{Z_1, \ldots, Z_m\}=\{Y_1, \ldots, Y_m\}$.
4) M imputes at least one feature in Y to T.

Let us unpack this in more detail. I will use an example from Frigg and Nguyen (2021) to do so. Frigg and Nguyen (2021) consider a car's wing mirror as analogous to scientific representation. By using a car's wing mirror one can reason about objects reflected in the mirror. In other words, the model M allows for surrogative reasoning about T—the target system (in this case how close or far a car is behind you, where M is the wing mirror). A wing mirror, however, is convex. A convex mirror makes objects appear smaller than they in fact are. The model M is a *Z-representation.* A Z-representation is the same as the *p*-representation discussed in the previous section. A Z-representation exemplifies M as having certain features, $Z_1, \ldots, Z_m$. Interpreting the exemplifying features of M requires a key—a way of potentially translating the exemplifying features of M into those features being representations of T as having those features. In this example, the relevant key is a sticker on the wing mirror that informs the driver that

"Objects in mirror are closer than they appear" (2021, 2439). Here is the core part that the DEKI account shares with the nonliteral view of idealization. One should not necessarily treat the imputation of the set of features *Z* to *T* literally. That is, if one judges the car in the wing mirror to be far behind, one would be mistaken. It is false that the car in the wing mirror is as far behind as it appears to be. However, this false interpretation arises only if *M* and its exemplifying features are treated literally. In other words, to reap the benefits of how the wing mirror represents the world, it must be treated nonliterally. This is both useful and legitimate as it provides or facilitates a grasp on the world outside the car. *Keying-up* is what allows for a nonliteral interpretation of the reflection provided by the wing mirror.

The same result holds in case of *idealization*. Like Mäki (2020), Frigg and Nguyen take a literal interpretation of idealizations to be problematic. They say:

> Just because a model exemplifies an infinite number of individuals, this feature does not have to be interpreted as holding in the actual population. Just because the idealised pendulum exemplifies being subject to no air resistance, it needn't be taken to represent the actual pendulum in the grandfather clock in your office as being subject to no air resistance or friction. (Frigg and Nguyen 2021, 2440)

Mäki (2020), the DEKI account of Frigg and Nguyen (2021), and the nonliteral view of idealization presented here thus converge on the same point: it is a mistake to treat idealizations as deliberate falsehoods—as false assertions about the world (contra all those identified as endorsing the FvI in chapter 2). It is theoretically misleading to interpret idealizations as being representations of *p*—as a set of features that represents the actual properties of target systems. The nonliteral view of idealization is, therefore, not novel in being the only nonliteral view of idealization. I take this to be a strength of the account. It lends support to a nonliteral treatment of idealization that several researchers are converging on similar points. Moreover, it shows that several researchers are working to formulate the correct view of idealization—a view that runs counter to the FvI.

However, unlike Frigg (2010), I do not draw the conclusion that scientific models are fictions (see chapter 10). According to Frigg (2010), the core of the fiction view of models is the claim that scientific models are similar to the places and characters one finds in literary fiction. For example, when modeling phenotypic distribution as involving no mutation, drift, and infinite populations, biologists are describing an imaginary population.

According to Frigg: "These imaginary scenarios are tellingly like the places and characters in works of fiction like Madame Bovary and Sherlock Holmes. These are scenarios we can talk about and make claims about, yet they don't exist" (2010, 101). The nonliteral view of idealization does not deny there are close links between idealized posits and fictional characters. However, unlike Frigg (2010), the nonliteral view of idealization implies a version of realism about idealizations. They exist as part of our modeling practices in science, in the sense that idealized posits can be given precise mathematical formulations. They exist, as Thompson-Jones (2020) puts it, as a class of abstract artifacts even if there are no concrete and actual infinite populations, point objects, or simple pendula.

3.5 Concluding Remarks

This chapter has focused on the nature and function of idealization in scientific models, and how models that trade in idealizations relate to the world. I have argued against what is a widespread and pervasive view of idealization in the literature: that idealizations are useful legitimate falsehoods, which I called the "falsehood view of idealization" (FvI). The central argument is: The FvI rests on the unnecessary assumption that scientific discourse must be taken literally. However, as I have argued in this chapter, it is a mistake to take scientific discourse about idealizations at face value, since idealizations do not literally represent things in the world as being thus and so.

4 Idealization and Scientific Realism

Overview: The aim of this chapter is to apply the nonliteral view of idealization to the discussion about scientific realism. The general shape of the argument will be: idealizations only pose a serious challenge to scientific realism insofar as they are treated as truth-apt assertions about the world. Hence, only the FvI represents a challenge to scientific realism. By contrast, the nonliteral view of idealization can be used as a defense of scientific realism by vindicating the view that the aim of science is to provide us with (approximate) truths about the world.

4.1 Introduction

Consider the following two paragraphs from Godfrey-Smith's *Theory and Reality*: What does science try to describe?

> The world, of course. Which world is that? Our world, the world we all live in and interact with. Unless we have made some very surprising mistakes in our current science, the world we now live in is a world of electrons, chemical elements, and genes, among other things. Was the world of one thousand years ago a world of electrons, chemical elements, and genes? Yes, although nobody knew it back then.
>
> But the concept of an electron is the product of debates and experiments that took place in a specific historical context. If someone said the word "electron" in 1000 A.D., it would have meant nothing—or at least certainly not what it means now. So how can we say that the world of 1000 A.D. was a world of electrons? We cannot; we must instead regard the existence of electrons as dependent on our conceptualization of the world. (Godfrey-Smith 2003, 173)

The position defended in this chapter captures the general attitude expressed in the first paragraph: the world has a mind-independent structure, and one aim of science is to truthfully or accurately (or, approximately truthfully) describe that mind-independent structure. This position is known as *scientific realism*. I will not be claiming that science always succeeds in achieving this aim. A common way of describing scientific realism is by noting that it incorporates three theses (or stances). Psillos provides what many take to be the standard formulation of scientific realism:

1. *The metaphysical stance* asserts that the world has a definite and mind-independent natural-kind structure.
2. *The semantic stance* takes scientific theories at face-value, seeing them as truth-conditioned descriptions of their intended domain, both observable and unobservable. Hence, they are capable of being true or false. Theoretical assertions are not reducible to claims about the behaviour of observables, nor are they merely instrumental devices for establishing connections between observables. The theoretical terms featuring in theories have putative factual reference. So, if scientific theories are true, the unobservable entities they posit populate the world.
3. *The epistemic stance* regards mature and predictively successful scientific theories as well-confirmed and approximately true of the world. So, the entities posited by them, or, at any rate, entities very similar to those posited, do inhabit the world. (Psillos 1999, xvii)

Whether the mind-independent structure of reality has a natural-kind structure is debatable. I will not discuss this any further. The semantic thesis asserts that all scientific discourse is truth-valued, including claims about unobservable entities. Its main function is to fend off positions that threaten the realists' positive epistemic attitude toward scientific claims about unobservable entities (Psillos 1999). This claim is incorrect insofar as one accepts the nonliteral view of idealization presented in chapter 3. The realist view on science to be defended in this chapter rejects what we might call *unrestricted semantic realism*.

Unrestricted semantic realism is the claim that all scientific claims are truth-conditioned. In contrast to unrestricted semantic realism, the nonliteral view of idealization implies a form of *restricted semantic realism*: this is the view that only some scientific discourse should be taken as truth-apt, since claims about idealized posits do not function as representations of features in real-world target systems. If we take talk about idealizations seriously in science, which we should, then it follows that not all scientific claims are truth-apt.

The aim of this chapter is to defend scientific realism in the context of idealized model-based science. The reason is that idealization has been argued to present the greatest challenge to scientific realism (see, e.g., Chakravartty 2001). The idea is easy to motivate: most scientific realists think that a reasonable aim of science is to accurately or truthfully describe what reality is like, and where "truth" is usually grounded by appeal to similarity or correspondence relations between a scientific model and its target system. However, idealizations in science will ultimately fail any such test for similarity and correspondence precisely because they do not represent anything that exists in any target system. I argue that it is only the FvI (see chapter 2) that causes a problem for scientific realism. This is because the FvI makes controversial the realists' positive epistemic attitude toward scientific claims being frequently approximately true about unobservable entities.

The standard formulation of the semantic thesis is a literalist thesis about the nature of scientific discourse—namely, that all scientific claims must be either true or false. However, and if interpreted literally, the FvI implies that scientific models, even if explanatorily successful, will turn out to be frequently false in virtue of the fact that they are partially composed of idealizations—understood as useful and legitimate falsehoods. This is especially the case whenever the idealizations in question are deemed to be indispensable. Because the FvI has this implication, it presents a problem for the epistemic thesis of scientific realism because it will turn out that the predictive success of well-confirmed and approximately true models of the world does not rest on entities that actually inhabit the world.

To see this more clearly, note that scientific realism is usually formulated as a veritistic thesis about the aim of science, with the added claim that scientific discourse should be construed literally. Veritism is the view that truth is necessary for epistemic acceptability. If scientific realism is construed as having a veritistic epistemology, then the ultimate aim of science is the discovery of truths about the world. Or, as Putnam (1975) contends: the success of science would be a miracle were our best theories not (approximately) true. Call this the *standard epistemic success argument* for scientific realism. However, given the ubiquity of idealizations, laws, models, and approximations, all of which are accepted as not true but nevertheless critical to the explanatory success of science, the claim that mature and predictively successful theories are (approximately) true is false, if taken literally. The epistemic stance of scientific realism looks problematic in this light. Thus, the standard epistemic success argument for scientific realism is problematic. In this chapter, I argue that model-based science, although ripe with idealizations, is consistent with scientific realism, but only if some scientific discourse is taken nonliterally.

Here is how the argument sits in the current literature. Consider first the relation between scientific realism and idealization. Suárez (2010) focuses on idealization cast in terms of fictionalism. He is right to point out that it is often assumed that fictionalism is in opposition to scientific realism. Suárez concludes that the status of fictionalism in the burgeoning literature on scientific representation is neutral in the realism vs. antirealism debate. Elliott-Graves and Weisberg (2014) draw the same conclusion. Frigg and Nguyen (2021) agree, although they frame their discussion in terms of keeping intact a veritistic understanding of scientific theories, while denying that scientific claims must be taken literally. I depart from these authors in the following way: the claim that idealization is neutral between realism and antirealism concedes too much to the antirealist. Or, this concession does not do enough for the realist. It does not do enough to block the (alleged) appeal of scientific antirealism.

In contrast, Strevens's (2017, 2019) discussion of idealization aligns with realism about science. He argues that all idealizations work in the same way: they show that some actual feature of a phenomenon is irrelevant to the explanation of that phenomenon. This is true, but only in some cases (see chapter 3). Unlike Strevens (2008, 2017, 2019), I return to some familiar territory in the debate; namely, that some nonactual features of scientific models are explanatorily relevant for the target phenomenon (see also Shech 2018).

In motivating this claim, I adopt a view of (most) scientific models as consisting of two types of claims: (a) idealizing and noncorresponding claims; and (b) nonidealizing and corresponding claims. One way to look at the distinction between type (a) and type (b) claims is that this way of depicting scientific models is quite common in the literature. Indeed, most distinguish between nonidealizing claims and idealizing claims. Similarly, most scientific models will distinguish between corresponding and noncorresponding claims. Here is the key difference. The assumption that idealizations misrepresent aspects of target systems implies type (a) claims. Therefore, those working with the FvI (see chapters 2–3), assume that idealizations are falsehoods because they fail to correspond to aspects of target systems. The nonliteral view of idealization developed in chapter 3 interprets the nature of type (a) claims differently: type (a) claims are idealizing and noncorresponding because the function of idealizations is not to represent aspects of target systems. Hence, the way I understand type (a) claims can be captured in the form of type (a*): a scientific model is partly composed of a type (a*) claim in as much as idealizations do not serve any representational function from a model to a target system. I will leave the use of (a*) aside for the rest of this book. Just keep in mind that when I endorse the view that scientific models are composed of type (a) and type (b) claims, I treat type (a) claims nonliterally.

I proceed as follows. In section 4.2, I start by looking at the central dilemma for scientific realism as it is formulated by Psillos (2011). I show that the only way to preserve a veritistic picture of the aim of science is to argue that Psillos is wrong in seeing this as a dilemma, when there is a third option available for the realist about science. In section 4.3, I move on to consider several arguments that problematize veritism in science. I show that none of these arguments work because they all presuppose the FvI. In section 4.4, I start my defense of the standard epistemic success argument for scientific realism. I do this initially by looking at the relation between idealization and scientific realism. I argue that idealization allows for a defense of veritism and I do this by arguing that type (a) claims should be treated nonliterally, while type (b) claims can be given a literal treatment. Crucially, nonliteralism about type (a) claims retains the epistemic stance of scientific realism even in the context of idealization. In section 4.5, I consider the specific counterargument that the position I develop in this chapter is better understood as a version of instrumentalism (a version of scientific antirealism). I conclude in section 4.6.

4.2 The Central Dilemma

Consider Psillos's (2011) central dilemma: either scientific theories should be understood literally, which makes them false, or scientific theories should be understood nonliterally, but then scientific realism would be false (given the semantic thesis of scientific realism). The central dilemma partially captures the problem for scientific realism given the substantial role played by idealizations in model-based science. On the one hand, if we take scientific theories (qua families of models) at face value—as part of a truth-conditional view of scientific discourse—then scientific realism is problematic given the ubiquitous presence of idealization in scientific modeling. On the other hand, if we pursue a nonliteral understanding of scientific discourse, then scientific realism will be unstable (given the literalist basis of the semantic thesis for scientific realism).

Let us assess the status of the central dilemma for scientific realism. First, it fails to capture that there is a third option for scientific realism. The choice is not between complete literalism and complete nonliteralism. It is entirely possible to argue that some scientific discourse should be taken at face value, while other kinds of scientific discourse should not be taken at face value. Second, the real problem rests with the first horn of the dilemma, which is the one Psillos (2011) opts to defend. The solution to the dilemma is, according to Psillos, to defend literalism about the semantic thesis and accept that idealized posits are ontologically real. Psillos says:

> A literal understanding of scientific theories implies commitments to models qua abstract objects. Insofar as these theories (taken literally) are true, there are these abstract entities. . . . They are abstract objects that can stand in representation relations to worldly systems. (Psillos 2011, 8)

Psillos's (2011) main claim is that scientific theories can be understood literally and still be true, although the price to pay is that the truth of scientific theories goes through the models of which the theories are composed (the semantic view of scientific theories). Psillos considers the linear harmonic oscillator. It consists of a mass term m, where a force F pulls the mass in the trajectory of a point $x=0$, and relies only on the position x of the body and a constant k. Following Giere (1988), Psillos asks us to suppose that the linear harmonic oscillator is introduced by means of a theoretical definition; namely, that a linear harmonic oscillator is whatever entity satisfies –, where the definiens is: the motion (along x axis) satisfies the following second-order differential equation:

$$F=md^2x/dt^2=-kx.$$

Idealization and abstraction have both played essential roles in arriving at this equality. Here is the point Psillos wants to make:

> It has been assumed that the angle of swing is small, that the velocity-dependent frictional force (air resistance) is zero, etc. There is nothing in the physical world that satisfies this definition. And yet the definition *is* satisfied by some entity, viz., the linear harmonic oscillator. (Psillos 2011, 8)

The result, according to Psillos, is that if a literally interpreted theory is true, then the kinds of entities needed for its truth are ontologically real. I take this to be controversial. To see this, let us foreshadow a discussion to come about the status and standing of mental or neural representation in chapter 5. I argue that neural representations are best understood as idealizations. Suppose we have a theory of cognition that makes use of theoretical terms such as neural representations. Let us also assume that this theory is true, or approximately true. Does it follow that neural representations are ontologically real? No. They may still be only abstract, idealized explanatory devices. This does not imply that appeals to idealized constructs, including appeals to neural representations, cannot be explanatorily useful. For instance, Chirimuuta (2024) argues that the "appeal to neural representation allows for an important simplification in cognitive neuroscience because it facilitates and justifies the black-boxing of intermediaries between brain responses and the distal objects that are relevant to the explanation of behavior" (160). Here "neural representation" is part of an explanatory strategy. It simplifies an overly complex explanatory task. One plausible reading is that even if we treat discourse about neural representation nonliterally, we might still take it seriously. However, as Levy (2011) puts it in a discussion about informational discourse in biology, we should not take such talk "too seriously" (652). If neural representation is an idealization, then it is not true that brains instantiate patterns with representational properties. The result is that, for explanatory purposes, one can treat "neural representation" as one would treat "point object" in physics: as an idealized (imaginary, nonactual) element suited for certain kinds of explanation.

I set this issue aside for the remainder of this chapter. Opting for the first horn of the dilemma will not suffice to justify the semantic thesis in light of the FvI. According to Psillos, scientific theories acquire their representational content via their models of target systems. Scientific models are abstract entities in the sense just described. Regardless of whether we adopt a deflationary or inflationary attitude about the ontology of abstract entities, scientific models will not represent target systems truthfully if scientific discourse is treated literally. Indeed, as Psillos notes, there is nothing in the world that satisfies many of the assumptions of the linear harmonic oscillator. If the FvI is in place, which it is for Psillos, then the epistemic thesis of scientific realism becomes problematic to defend.

The proper response for the scientific realist is to point out that some scientific discourse can be taken nonliterally—discourse involving idealizations. This formulation of the semantic thesis supports the epistemic thesis. It supports the epistemic thesis because it facilitates the possibility of true or approximately true explanations of target systems—even if such explanation comes from models that themselves are not strictly speaking true. What Psillos (2011) fails to appreciate is that even if one endorses the reality of abstract entities qua scientific models, scientific models will keep serving up falsehoods more frequently than truths, given the commitment to literalism about scientific discourse (see also Rowbottom 2022 for a similar kind of worry). Hence, to secure the epistemic thesis about scientific discourse, neither the first nor the second horn of the central dilemma will work. The only option, I submit, is to endorse a restricted form of semantic realism.

4.3 Motivating the Problem for Veritism in Science

In *True Enough* (2017), Elgin presents a series of arguments for why truth cannot be, and should not be, the epistemic aim of science. The basic structure of the argument is that veritism cannot be defended as the paradigm for the epistemic success of science if much of what is produced in science is not true, even if epistemically acceptable. Hence, epistemic acceptability of scientific claims can come in the absence of truth. According to Elgin:

> Understanding is often couched in and conveyed by symbols that are not true. Many of them do not even purport to be true. Such symbols are epistemically *felicitous falsehoods*. (Elgin 2017, 23; emphasis added)

Science deals with equations (e.g., the ideal gas law, the HH model), diagrams (e.g., depictions of harmonic oscillators), simulations (e.g., computational models of hemineglect), idealizations (e.g., infinite populations in biology, cellular automata), and many other features that would be, strictly speaking, false if taken to provide a literal interpretation of their referents in the world (e.g., infinite particles in my boiling kettle). There are no infinite populations, no natural gases that exhibit zero molecular interaction, and no boiling kettles with infinite particles. According to Elgin, models containing idealizing claims are false with respect to their target systems. However, despite being false relative to their actual world, target systems, it is their inaccuracy that "afford epistemic access to their objects" (2017, 23). Hence, they are useful (felicitous) falsehoods. Potochnik develops a similar line of argument: "If false posits or fictions facilitate at least some varieties of scientific explanations, this role is another form of sacrificing some truth to facilitate other aims" (2023, 153). Idealizations, or other kinds of modeling techniques in science, cannot be accommodated by a veritistic view of the epistemic aims of science. Khalifa sets out the general structure for this kind of argument in the following way:

P1: If veritism is true, then feature *X* of understanding is not epistemically valuable.

P2: *X* is epistemically valuable.

Therefore: Veritism is false. (Khalifa 2020, 945)

Elgin (2017, 27) considers an ideal pendulum, which is a rigid body consisting of a small mass oscillating over an angle at the end of a string. An ideal pendulum model of this kind provides a model for simple harmonic motion. Yet, it deviates from reality by idealizing away friction and gravity, and it sets the value of the mass of the string to zero.

Explanations that represent the pendulum as a simple harmonic oscillator or that model natural gases as if their values for molecular size and intermolecular collisions were zero "would be epistemically unacceptable if abject fidelity to truth were required. . . . If veritism is correct, its falsity makes it epistemically unacceptable. But scientists find it unexceptionable" (Elgin 2017, 27). In fact, the ideal gas law, despite containing falsehoods about reality, has substantial real-life applications. It has instrumental epistemic value even if it is a falsehood. Its instrumental value can therefore not be accounted for by veritism.

One might try to push back against this argument by pointing out that the standard epistemic argument for the success of science requires only that theories are partly true, or approximately true. This is a standard view (Papineau 2010). However, as Elgin argues, "not all felicitous falsehoods are even approximately true" (2017, 30). Elgin refers to the Hardy-Weinberg model in population genetics as a case in point. According to Elgin (2017), scientific models of this kind, of which there are many, are not even approximately true of their target systems in the world; they are false. However, they are epistemically acceptable because they prove to be instrumentally epistemically valuable in understanding the phenomenon under investigation. This is the problem with the standard epistemic argument for the success of science based on a veritistic epistemology. If idealized models facilitate understanding of reality, and the consensus is that many such models do, it cannot be in virtue of being truth-assertive about reality. This undercuts a defense of scientific realism, at least when taken to be about the discovery of the real structure of the world based on approximately true models.

4.4 Idealization and Scientific Realism

Thus far I have attempted to motivate two issues. The first was to highlight that the central dilemma for scientific realism is no dilemma, since a third option is open for scientific realism to treat at least some scientific discourse nonliterally. The second was to highlight a set of independent reasons for problematizing veritism based on the ubiquitous employment of useful falsehoods in science, which, despite their status as falsehoods, are epistemically acceptable. It is time to tackle head on the defense of veritism.

I started with the observation that philosophical discussions of scientific realism are formulated, for the most part, on a veritistic epistemology. Yet, as Suárez (2010) points out, many take this insistence on truth for scientific understanding to be incompatible with the further claim that scientific theories are not true when their claims are taken literally. Elgin (2017) falls into this group. Frigg and Nguyen (2021) turn this into a paradox between the following three propositions:

1. truth is necessary for epistemic acceptability (veritism);
2. scientific claims have to be taken literally (literalism); and
3. much of what model-based science produces is not literally true even if epistemically acceptable.

(3) is uncontroversial, since most agree that science is shot through with models employing strategies such as simplification, idealization, and approximation. However, if (3) is accepted, then (1) and (2) cannot both

be true. Whereas Elgin (2017) and Potochnik (2017) deny (1), Frigg and Nguyen (2021) deny (2) and defend (1) and (3). I also defend (1) and (3). I depart from Frigg and Nguyen (2021) in the following way: they argue that denying (2) enables one to accommodate (1) given (3). Yet, they think this cuts across the distinction between realism and antirealism. In other words, a denial of (2) can accommodate "realists and anti-realists positions alike" (2021, 2442). As I said in the introduction to this chapter, this concedes too much to the antirealist precisely because idealization in science has been deemed the "greatest challenge to the realist" (Chakravartty 2001, 329). My aim in this section is to defend (1) and (3) as a promising way of arguing for scientific realism. Let us start by looking at (2) when considering the standard epistemic success argument for scientific realism.

4.4.1 The Problem with Literalism

The view that scientific discourse must be taken literally rests on a fallacy; what I shall call the *literalist fallacy*, building on recent discussions in other work (Kirchhoff, Kiverstein, and Robertson forthcoming):

- Premise 1: Scientific models introduce distortions into how they represent target systems via idealization.
- Premise 2: The standard epistemic success argument for scientific realism requires that assertions about target systems must be literally true.
- Premise 3: Scientific models are not literally true or literally accurate representations of their target systems.
- *Therefore*, the standard epistemic success argument for scientific realism is false.

This kind of argument is problematic because it assumes literalism. Since scientific models are not literally true of their target systems, the standard epistemic success argument for scientific realism is controversial. This conclusion does not follow, because premise 2 is false. The central reason for denying (2) is its literalism about scientific claims. For idealizations to be any kind of direct threat to scientific realism, they need to come out as falsehoods, since this is the outcome that jeopardizes (1)—veritism. However, for idealizations to come out as falsehoods, they must be given a literalist reading. Yet, giving idealizations a literal interpretation is an instance of committing the literalist fallacy, as outlined above. A nonliteral treatment of idealizations results in them not being false (following on from chapter 3). Hence, (1) is compatible with (3). A denial of (2) brings with it another positive for the scientific realist: When scientific theories and models yield true or approximately true explanations, this capacity does not rest on a set of false assumptions about the world.

I hope to have said enough to motivate why (2) can be questioned. The next subsection defends (1) and (3) as a way of understanding the standard epistemic success argument for scientific realism.

4.4.2 Truth, Idealization, and Scientific Realism

Lawler provides a useful distinction to get us started. She identifies and distinguishes between two kinds of questions:

- [The] relation question: How do the falsehoods (or the models that involve them) *relate to* the phenomenon in question?
- [The] content question: How do the falsehoods (or the models that involve them) *figure into the content* of understanding the phenomenon? (Lawler 2021, 6864–65)

Lawler notes that the content question matters for the factive vs. nonfactive discussion about understanding, whereas the relation question, "taken by itself, does not" (2021, 6865). The issue of concern in this chapter is how best to address the relation question, where this will involve addressing if, and how, idealizations represent the target phenomenon under investigation. The relation question is at the heart of the discussion over scientific realism: What is the relationship between scientific models involving idealizations and the realities these models purport to represent? This is the map problem I introduced in chapter 1. I will proceed by way of examples.

4.4.2.1 The ideal gas law Consider again the ideal gas law: $PV=nRT$. According to Strevens (2008), idealizations of this kind say something true and false with respect to their explanandum. It is true that gases are made of fast-moving particles. It is also true that the particles expand to fill their containers uniformly. However, some of what the law says is false, if interpreted literally. It is not true of natural gases that the particles are infinitely small. It is also false that the particles do not collide, as represented in the model. From the perspective of the nonliteral view of idealization, a natural way to view Strevens's (2008, 2017, 2019) account of idealization is that by flagging certain aspects of target systems as explanatorily irrelevant, it thereby highlights that the idealized posits are not being used to represent the corresponding target system. Yet, this is not the position on idealization adopted by Strevens. As he says: "Many explanatory models contain idealizations, that is, deliberate falsifications" (2017, 1). Or, as Strevens states:

> On the causal difference-making approach, an important part of explanation is to draw a clear line between difference-makers and nondifference-makers. Idealization accomplishes this task in a showy way: it falsifies certain nondifference-makers, by setting a variable that represents them to an extreme or default value. (Strevens 2017, 4)

I adopt a different interpretation of idealization, as presented in chapter 3. Not all, but only some of the features of a model function to represent aspects of target systems. A model that conforms to *S*, where *S* takes the form "*p* and p_1 . . . and p_n and *J*," highlights that *p* and p_1 . . . and p_n serve to accurately represent aspects of target systems, while *J* is invoked in a way that does not imply that it misrepresents its target system. As it turns out, this is a more natural way of interpreting what Strevens's (2017) account of idealization implies. That is, by setting the value of a particular model parameter to zero, one is thereby not saying that the particles do not interact in target systems. However, if this is correct, then setting the value of some parameter to zero is not the same as saying that this parameter is false in the sense that it functions as a misrepresentation. Rather, setting a parameter to zero makes it salient that it is not explanatorily relevant for the purposes of explaining the Boylean curve of a gas, for example. This allows one to treat instances of *J* nonliterally. Or, the causal difference-makers of the model can be given a literal treatment, whilst causal non-difference-makers can be treated nonliterally.

Does the nonliteral view of idealization help to save veritism? Does it help with a defense of the standard epistemic success argument for scientific realism? I think it does. A key reason for this is that causal difference-making accounts build in accuracy requirements for scientific models to explain target systems. As Strevens says: "No causal account of explanation—certainly not the kairetic account—allows nonveridical models to explain" (2008, 297). The ideal gas law explains the behavior of target systems, given certain conditions, because "the overlap between an idealized model and reality . . . is a standalone set of difference-makers for the target" (2008, 318). Treating the idealized factors nonliterally enables idealized models to provide an accurate representation of relevant causal difference-makers in the target system of interest. However, if *S* is treated literally, then *S* is false because *J* is false. The nonliteral view of idealization blocks this outcome.

Here is where I depart from Strevens's account. Strevens claims that all idealizations work in the same way: "An idealization does not assert, as it appears to, that some nonactual factor is relevant to the explanandum; rather, it asserts that some actual factor is irrelevant" (2008, 316). But this generalization does not apply to all cases of idealization. Several cases discussed in the literature highlight idealizations that are indispensable for explaining certain target phenomena—for example, using the thermodynamic limit to explain the macroscopic phenomenon of phase transitions. The thermodynamic limit is a nonactual factor that exists on the side of the scientific model that help explain the occurrence of phase transitions on the side of the target system (see, e.g., Batterman and Rice 2014; Shech 2018; also discussed by Lawler 2021).

I now turn to consider limit functions more broadly.

4.4.2.2 Limit functions An interesting class of idealizing models is mathematical models that involve limit functions. One can ask: Did your weight change by zero kilograms just now? Any ordinary scale will inform you that there has been zero change. However, if one were to perform an atomic measurement, your weight change would be nonzero; some mass will have been lost or gained—for example, through respiration, osmosis, and so on. Take a different situation. Say you drop an ice cube in a glass of water at room temperature. If you measure the water temperature over time, it will approach room temperature. Here is a further case. Engineers face a specific challenge when having to determine the fuel flow through an engine. To simplify the measurement, they often divide the flow curve into smaller units that are easier to measure. In all these cases, a limit function defines the value that a function approaches as an input variable gets arbitrarily close to a particular point. Limit functions are more commonly used in today's differential calculus. Both the limit function and the infinitesimal allow for the quantification of uncertainty, and therefore, allow optimization of certainty about changes in target systems, especially in explanations of motion, including trajectories, movements, and velocities.

These cases are all examples of certain patterns of behavior that one might wish a theory to explain—the target system. The explanatory task then becomes the construction of a model that allows for an explanation of that system's behavior. When engineers use limit functions to calculate fuel flow in an engine, their mathematical posits are idealized by representing the target system geometrically to derive the fuel flow by dividing the curve over the flow into smaller units that are easier to measure. In this sense, the model involves idealizing, noncorresponding claims that are neither true nor false with respect to the target system. However, when limit functions are used in this way, they facilitate the production of nonidealized, corresponding claims about the rates of change in the target system. Here we have an example of a mathematical model that includes idealizing, noncorresponding claims and nonidealizing, corresponding claims. If treated nonliterally, the model turns out to be approximately true of its target system. The central point to keep in mind is: idealized posits are not false of the world if given a nonliteral treatment. Moreover, if only considered in relation to the model they are a part of, type (a) claims can be treated literally. That is, one can give a literal treatment of the mathematics involved in limit functions, while giving a nonliteral treatment of the same mathematical formalisms across the model-world relation. Hence, if scientific models contain type (a) and (b) claims, one does not need to conclude that type (a) claims are false of the world they represent, because their function is not to truthfully or accurately represent the world.

This outcome also holds for many mathematical and computational models in cognitive science. We can put this in terms of a problem for certain parts of naturalistic philosophy of cognitive science. This is what Chirimuuta (2024) does, and I find it entirely correct. She states: "To say that the cortex is literally computing the function stated in a model is . . . to identify two fundamentally different things" (245). For example, in chapter 6, I discuss a model of early vision that includes the function of computing the Laplacian of a Gaussian distribution. The function can be used to model processes in early vision. Yet, it does not follow that neural systems literally compute this specific function. Those inferring otherwise commit what Chirimuuta calls "the fallacy of misplaced concreteness." This is the fallacy of "taking the abstractions of science for concrete reality, confusing the model with the target" (2024, 246). What we need to be attentive to is that certain theoretical constructs in scientific models can be explanatorily useful even if they are not part of actual target systems. Chapter 9 explores this in detail with a case study from theoretical neuroscience.

4.5 Objection: The Nonliteral View Is Really a Version of Instrumentalism

Let us start by making a distinction between two kinds of scientific realism. First, there is *individual* scientific realism, which asserts that mature and predictively successful theories are (approximately) true. Second, there is *selective* scientific realism, which asserts that only the stable and working parts of mature and predictively successful theories are true. According to Campbell: "Selective realism is realistic realism, that is, realism more true to scientific life than any blanket, undiscriminating doctrine" (1994, 27).

One might try to argue that the problem of idealization (the one arising from the conjunction of model descriptions with idealizing assumptions) does not arise for selective realists. Prima facie, at least, it would seem that the version of scientific realism explored and defended in this chapter is a version of selective scientific realism. Selective realism is usually seen as one of the most popular ways of responding to the *pessimistic meta-induction* (PMI) argument against scientific realism. The PMI argument states that because foundational ideas have shifted so often across scientific theories in the past one should expect our current theories to turn out wrong (see, e.g., Laudan 1977; Mach 1911; Poincaré 1952; Putnam 1978). Selective scientific realism can be seen as a way of overcoming the PMI argument, since selective scientific realism is committed to the claim that only the stable, working parts of a theory are carried across from one theory to the next. So, selective realism commits only to the stable, working parts of a theory and suspends judgment about so-called idle posits. Crucially, selective scientific realism takes the "working parts" to be the true parts of the scientific theory.

We have already seen an example of this in Strevens's account of idealization. As we know, for Strevens (2017), idealizations function to flag that some elements of the target system are causal non-difference-makers—such as the size of the molecules and their interaction. These two causal aspects make no explanatory difference to deriving the behavior of a gas based on the ideal gas law (Strevens 2017, 2019). The causal-difference makers are the "working parts," whereas the idealizations are the "idle parts." This would be an example of selective realism.

However, selective scientific realism is not immune to the problem of idealization raised in this chapter. The reason is that the seemingly simple solution of identifying idealizations as idle parts and only causal-difference-makers as working parts is not available for selective realists when those idealizations are indispensable for deriving successful predictions (i.e., when the idealization belongs to the working parts). That is, if idealizations are indispensable for explanatory success, selective realists cannot maintain that the truth of those parts (including idealization) is the best explanation of success in science. Therefore, the problem of idealization arises for selective scientific realism.

The version of scientific realism developed in this chapter need not be seen as a version of selective scientific realism. The reason is that I have made the claim that for at least some scientific models, it is the idealizations that ground their explanatory accuracy. The idealizations in the ideal gas law are an example, since they form part of the derivation base of the ideal gas law (see also Lawler 2021). That is, the idealizations are not peripheral to the ideal gas law; they are indispensable for its formulation. We will see the same in chapter 9, where I discuss the role of variational techniques in theoretical neuroscience and the variational free energy formulation for modeling how organisms are able to keep an upper bound on entropy.

We can now state the problem for the nonliteral view of idealization. To get out of the problem of idealization for scientific realism, it will not be sufficient to say that the relevant idealizations are not false but instead are not truth valued. The core of the problem is that the position defended in this chapter is committed to the claim that some of the working parts of a model are not true. So, just as for selective scientific realism, the problem for the nonliteral view of idealization is that it cannot maintain that it is the truth, or approximate truth, of the working parts that is the best explanation of the explanatory success of a model using idealizations. But the implication of allowing non-truth-valued assumptions to play central roles in the explanatory success of science is a slippery road to a version of antirealism called *instrumentalism* (many thanks to one of the anonymous reviewers for raising this objection).

This objection rests on a problematic premise—namely, that the link between a scientific model and its target system that explains the success or accuracy of the model is that the model itself is true or close to being true. However, accurate representation between a model and its target system is not the only way by which to establish a link between a model and a target system. Godfrey-Smith (2003) has developed a version of this view. The key insight that he provides against this kind of objection to scientific realism is that realists "need to give up the idea that success in science points directly or unambiguously toward the truth of theories" (2003, 178). This assessment of the situation is shared by the nonliteral view of idealization, which is why I have argued that idealizations in scientific models have various functions—although none of them is to be truth-apt.

Is this enough to block the threat of instrumentalism? In chapter 3, I made use of a particular example from Rowbottom (2022) to illustrate the nonliteral view of idealization—the sentence *S* with the form "p and p_1 . . . and p_n and *J*" (Rowbottom 2022, 884). *S* is the proposition that "electrons have mass and charge and spin." In his 2019 book, Rowbottom makes use of this example in a defense of instrumentalism. Here is what Rowbottom states:

> We now come to the question that inspired this section. Should we take talk of electrons literally? The property instrumentalist's answer lies in the negative. But talk of some tiny thing possessing discrete mass and charge may be taken seriously; and a property instrumentalist might accept that we have isolated such an entity, while believing that the talk of "spin" (or "intrinsic angular momentum") is just a way to account for its behaviour in some circumstances. It is crucial to note, however, that spin need not be understood to be a fiction merely for explaining how observables interrelate. Rather, it may be taken to be a fiction for predicting—and furnishing us with an understanding of how those unobservable things "similar to electrons" behave. (Rowbottom 2019, 54)

Scientific discourse about p and p_1 . . . and p_n can be treated literally. However, claims about *J* should be treated nonliterally. This is the position defended by Rowbottom (2019). It is also the position defended in chapter 3, where I developed the nonliteral view of idealization. So how could the nonliteral view of idealization not imply a version of instrumentalism? The right response to this kind of objection is to make use of the distinction I made earlier between *unrestricted* and *restricted* semantic realism. The correct solution is to adopt restricted semantic realism, a proposal that Rowbottom (2022) curiously argues is friendly to the agenda of scientific realism. The other response is to follow Godfrey-Smith (2003) in relaxing the link between explanatory success and the (approximate) truth of scientific models. As I see it, restricted semantic realism implies the kind of relaxed attitude advocated by Godfrey-Smith (2003). If this is correct, then the nonliteral view of idealization is entirely consistent with the version of scientific realism defended in this chapter. Or, differently put, the nonliteral view of idealization does not imply instrumentalism.

This specific result will prove important for the discussion in part II, when turning to apply the nonliteral view of idealization to the topics of neural representation, neural computation, and the prospects for explanatory unification in cognitive science. Scientific realism, in its standard form, asserts that reality has a mind-independent structure, and that scientific theories and models are to be interpreted literally, as truth-apt representations of their target domains. However, scientific models, including basically all models in cognitive science, introduce distortions into their way of modeling target systems via idealization and approximation. Hence, scientific models are not true and accurate representations of their respective target domains. Consequently, scientific realism about models in cognitive science cannot be defended. This conclusion rests on a fallacious inference: the fallacy of inferring the truth of a version of scientific antirealism based on the observation that the properties of models in cognitive science do not literally map onto target systems. The problem is literalism about scientific discourse. There is no scientific model that is exactly like its target system (chapter 1). This is true in any model-based discipline, including cognitive science. Consider an example that takes center stage in chapter 9: the free energy principle (FEP). The FEP is a model of brain structure and function accompanied by Bayesian mechanisms. In the philosophical literature, and elsewhere, some take the FEP, and relatedly predictive processing, to mean that brains are literally predicting and inferring the oncoming stream of sensory input. Clark (2016) states: "Creatures equipped with ever-active, predictive brains are, of course, already (quite literally) 'ahead of the game': as brains like that are . . . constantly guessing the ongoing stream of sensory input, including the inputs that should result from their own next actions and worldly interventions" (176–177). If taken literally, this claim is false. Neurons do not literally make inferences. Inferences involve propositions. Frege (1979) provides the following definition: "To make a judgement because we are cognisant of other truths as providing justification for it is known as *inferring*" (1979, 3). The FEP provides a model by which to interpret brain structure and function, as well as self-organizing processes more generally. So, the FEP, and its formalisms, do not literally map onto the nervous system. But this is no reason to infer that the FEP, or any computational or mathematical model in cognitive science, is best interpreted via the lens of instrumentalism. The lesson to be appreciated is that we frequently need to treat scientific terms and formalisms in a nonliteral way. Another example is the widespread view that brains are computational. If taken literally, this claim is also false. Computational tools are useful and legitimate in cognitive science. But one should not infer from this that nervous systems are computational. Models can be computational. Yet computational models do not imply that target systems are computational (chapters 6 and 7). That is how the story told in Part 1 of the book applies to cognitive science.

4.6 Conclusion

The result of the above analysis is that idealization does not undercut the epistemic stance of scientific realism. Recall, the epistemic stance states that predictively successful theories are true (or approximately true). I have argued that it is possible to retain the veritist view that truth is required for epistemic acceptability, whilst acknowledging that much of what science produces is not literally true, even if epistemically acceptable. The key move in the argument for why idealization does not threaten scientific realism is the denial of literalism about scientific discourse involving idealizations. It allows for an appreciation of why accurate representation from a model to its target system is not the only way for scientific models to be successful.

II Implications for Cognitive Science

It is now time to apply the nonliteral view of idealization, and how I used it to defend a version of scientific realism, to a select but foundational set of topics in the philosophy of cognitive science. I aim to investigate to what extent scientific realism can be defended across a range of foundational topics in the philosophy of cognitive science, given the widespread use of idealization in cognitive science.

Cognitive science, just as in physics, chemistry, and biology, is in large part a model-based discipline (chapter 1). Often, but not always, models are proposed to account for some phenomenon or target system. For a model to do this kind of work, those using the model must make specific assumptions about its intended target system. Theoretical frameworks provide a general perspective and often a set of (qualitative and quantitative) tools for building models of cognition. These models are simplified models and they make use of idealizations. Because models in cognitive science are simplified and idealized, the map problem resurfaces in the sciences of the mind.

Across the next six chapters, I consider three foundational topics in cognitive science, a particular and concrete case study. I finish by showing how the nonliteral view of idealization does not imply any kind of fictionalism about scientific discourse in cognitive science.

Representation and idealization in cognitive science The concept of mental representation has long been a foundational concept in the philosophy of mind and cognitive science. A central question in the discussion about mental representation is: What sort of ontological attitude should researchers have toward these theoretical posits? Answers to this question tend to oscillate between two answers. Researchers should adopt either a realist attitude, in the sense that realism about mental (or neural) representations commits one to the existence of mental (or neural) representations. According to the realist about mental representation, cognition involves internal content-bearing, representational states and processes, and cognitive theories are those that put forth representational states and processes in their attempt to explain cognition. This is also called the *representational theory of mind* (RTM). Or, researchers should adopt an eliminative attitude, whereby one seeks to show not only that mental representations do not exist—at least at certain scales of investigation—but also that theoretical discourse involving mental or neural representation adds no explanatory value. There is a third option, that researchers should adopt a kind of fictionalist attitude, where appeals to representations "should be treated as something more like useful fictions or theoretical devices that play some sort of pragmatic, heuristic role in our theorizing" (Ramsey 2020, 54). Egan (2020), discussed in chapter 6, defends a view of this kind. On Egan's account, neural representations do not exist. They exist only in the eye of the scientist. As Egan states: "Representations are distinguished from mere causal relays by the fact that they are assigned content by the interpretation function f_1, but since the content assignment is confined to the heuristic gloss, it might be argued that the phenomenon of interest—has indeed disappeared" (2020, 43). Neural representations functions as idealizations for Egan. In chapter 5, I defend a similar view. However, I shall argue that it is far from obvious that the appeal to neural representation is explanatorily valuable.

I shall also argue (in chapter 5) that many realists about neural representation make use of neural representational posits in a way that is similar to how scientists make use of idealizations. In the previous chapter, I cited Chirimuuta (2024) as endorsing the view that the use of theoretical posits such as neural representation primarily facilitates "explanatory simplicity." Cao (2022) and Shea (2020), both realists about the existence of neural representation, appeal to the alleged fact that the use of neural representational discourse increase "cognitive tractability." Simplicity and tractability are functions of idealization in science.

It is also striking to observe that what many researchers say about scientific models mirrors what many say about the concept of mental representation. Cognitive science is a model-based science. Its models idealize and approximate like all other scientific models. Hence, posits such as neural representation are part of abstract and idealized models use to explain aspects about the target systems under investigation (Coelho Mollo 2021). Scientific modeling asks us to imagine the world as different from how it actually is (Godfrey-Smith 2009). This is also the case in the discussion on neural representation. Here one finds models that are explicitly said to be idealizations (Shea 2020). So again, models that invoke neural representational discourse describe a hypothetical scenario—a view of the functional pattern of neural activity that is different from how it actually is.

In chapter 5, I seek to establish that cognitive models generated based on the RTM are best understood as idealizations. I frame my discussion in terms of the nonliteral view of idealization (chapters 3, 4). I argue that this assessment yields three options for thinking about representations qua idealizations: (1) representational discourse should be treated nonliterally, is epistemically useful, and cannot be eliminated; (2) representational discourse should be treated nonliterally, is epistemically useful, and can be eliminated; or (3) representational discourse should be treated nonliterally, is epistemically useful, but irrelevant for causal explanation.

All three options can be made consistent with a version of scientific realism. However, none of the three options allows for a defense of realism about neural representations. The theoretical construct of a neural representation has its place within a model of the mind, not in the target system itself. Once one acknowledges that these posits are theoretical constructs, much like point objects or ideal gases, one question is: Do such posits do the same work as point objects and ideal gases? My own take on the situation is as follows. Option (1) can be challenged. The field is currently seeing an increasing amount of work suggesting that appeals to neural representation are not useful and should therefore be eliminated (see, e.g., Hutto and Myin 2013). They are not useful because they do not facilitate genuine explanation. Option (3) runs counter to any realist defense of neural representations. That is, if they are explanatorily irrelevant, then these posits will not be part of the working parts of a theory of cognition. Option (2) is the most defensible of these options, when neural representations are treated as idealizations. One reason is that cognitive science has yet to reach the same degree of maturity and predictive success of other sciences such as physics and biology. In this sense, using neural representational constructs might be useful but only in the short term. As with idealizations, the appeal to neural representation can facilitate tractability: these posits offer a way of abstracting away from the extremely complex neuronal processes underlying cognitive behavior. So, the use of neural representation is a simplifying strategy, much like the use of idealization in other model-based sciences. However, as I mentioned in the Introduction, in the case of appealing to mental or neural representation, simplification does not come with the added theoretical benefit of providing true (or even partially true) explanation of the target system or behavior in question.

Computation and idealization in cognitive science Another major topic in the cognitive sciences, and its philosophy, is computation. If you think that computation is central to the foundations of modern cognitive science, you have entered the doors of computational cognitive science, where the aim is to explain cognition (broadly understood) through the construction of computational models that specify the mechanisms and processes involved in realizing human and nonhuman forms of cognition. Work in computational cognitive science shares all the features of scientific models discussed thus far: they are abstractions and idealizations, and they seek to represent their target systems indirectly and differ dramatically in how accurately they do so. A particular feature of models in computational neuroscience is that their idealizations are usually permanent just because of the complexity of nervous systems. Another feature of models in this branch of science is the use of mathematical formalism employed to make certain kinds of analytical problems solvable—or to make certain computational problems tractable. In chapter 2, I considered an example by Thompson-Jones (2005), where several mathematical derivations were invoked to facilitate a computational explanation of the acceleration and flight time of a cannonball. I noted that the mathematical derivations of the model under discussion were not being attributed to the actual target system. A few additional examples will help to bring out these points.

Your heart pumps a volume of blood roughly equal to the volume of a can of Coca Cola every four seconds. It is possible to calculate the amount of blood pumped by the heart each minute by using the following formula: Q [L/min] = *HR* [beats/min] × *SV* [mL/beat] (Hryvniak et al. 2021). The heart does its job independent of this computational procedure. Objects, we are prone to think, obey certain general physical laws. One can describe the rate of acceleration at which bodies move toward one another by appealing to Newton's laws of motion. For instance, Newton's law of gravitation states that any particle of matter attracts any other particle of matter with a force varying directly as the product of the masses and inversely as the square of the distance between them. In mathematical terms, $F = G(m_1, m_2)/R^2$, where F is the magnitude of the attractive force, G is the gravitational constant, and R is the square of the distance between the particles of matter. Objects fall as they do independent of the law of gravitation. Or, as Hutto puts the point: "No one in this day and age would be tempted to explain the behavior of a falling rock by suggesting that it, or any of its integral parts, is attuned to such laws or that it exhibits a mastery of them" (2005, 393). In both cases, it would be false to say that the entities in question literally do what they do in virtue of computing specific mathematical functions.

According to Egan (2018), computational theories of early vision seek to explain processes such as edge detection by modeling the process in terms of computing the Laplacian of a Gaussian distribution of the retinal array. Taking the Laplacian of an image highlights regions of rapid intensity change, which is useful for edge detection. This is possible because the Laplacian is the divergence of the gradient of a function. Because images with rapid intensity change are noisy, the Laplacian is usually applied once an image has been smoothed—for example, by a Gaussian smoothing filter. The Laplacian $L(x,y)$ of an image with pixel intensity values $I(x,y)$ can be computed by using the following equation (Fisher et al. 2003):

$$L(x,y) = \frac{\partial^2 I}{\partial x^2} + \frac{\partial^2 I}{\partial y^2}.$$

Why should one treat brains and cognition differently from how we treat pumping hearts and falling rocks? We can model the function of hearts and falling rocks by using computational and mathematical methods. We can also model edge detection in early vision by means of computational techniques from machine learning. However, we do not, or at least we should not, literally think that hearts and rocks compute a mathematical function to exhibit the behavior that they do. Why should we take it literally, then, that edge detection in early vision functions as it does in virtue of computing a mathematical function? In other words, if we deny literalism in cases of hearts and rocks, should we not also deny literalism about computational discourse about the brain and cognition, when literalism is invoked from models to target systems? This is not merely about equal treatment of hearts, rocks, and brains; it is about methodology. We understand systems in the world by using models of the world. However, naturally occurring systems need not have the properties of their respective scientific models.

Even though modeling in computational neuroscience is often appealed to in order to motivate or underpin the computational theory of cognition (CTC), computational modeling has, in principle, nothing to do with the CTC. Rather, computational tools such as quantitative mathematical models are introduced to solve the analytical intractability of most mathematical dynamical descriptions (see, e.g., Murphy 2012; Piccinini 2020). For example, Griffiths et al. (2012) note that humans have about 1,000 olfactory receptors and can recognize between 1,000 and 10,000 odors. Yet, modeling these variables in search of some arbitrary pattern of olfactory receptor activation is computationally intractable. Quite a bit of work in computational cognitive science is devoted to solving issues of this kind by finding tractable solutions to intractable problems.

The CTC represents a theoretically and philosophically motivated thesis (see, e.g., Cao 2018). It asserts, in its basic formulation, that the brain and cognition cannot only be usefully modeled with computational techniques but that the brain and cognition are computational. The CTC operates on at least two core assumptions. (1) Scientific discourse must be taken literally—the CTC about the brain and cognition is either true or false (literalism). (2) Truth is necessary for epistemic acceptability—only if true about the brain and cognition, should the CTC be accepted. As I mentioned above, evidence for the CTC comes predominantly from computational neuroscience, which relies on computational models to study the brain and cognition. However, computational neuroscience is a model-based field where, as we know, the use of abstraction and idealization is pervasive. Evidence for this is easy to find. Sterratt et al., write:

> To make progress in the computational modeling of biological systems, abstraction of the elements of the system will be required. Without abstraction, the resulting models will be too complex to understand and it will not be feasible to carry out the desired computations even using the fastest computers. (Sterratt et al. 2011, 316)

Sejnowski, Churchland, and Koch, make a similar point:

> Textbook examples in physics that admit exact solutions are typically unrealistic, but they are valuable because they illustrate physical principles. Minimal models that reproduce the essential properties of physical systems, such as phase transitions, are even more valuable. The study of simplifying models of the brain can

> provide a conceptual framework for isolating the basic computational problems and understanding the computational constraints that govern the design of the nervous system. (Sejnowski, Churchland, and Koch 1988, 1300)

Piccinini and Bahar write:

> When a computer program implements a mathematical model of a neural process, the computations of the simulating program are a means for generating approximate representations of the states that the mathematical model attributes to the neural process. This kind of model attempts to capture something essential to a neural process based on what is known about the process by employing suitable abstractions and idealizations. (Piccinini and Bahar 2013, 478)

Across chapters 6 and 7, I raise and discuss the following question: In what sense can one defend realism about the CTC? I argue that one cannot. However, as with my discussion on neural representations, I argue that computational approaches to the study of the mind can be shown to be consistent with a version of scientific realism. My reason for this conclusion is that the CTC is, generally speaking, defined in terms of state transitions—from input, internal state to output—governed by a rule or algorithm. I argue that the "rule-governing" aspect of the CTC is firmly located on the side of the scientific model rather than the target system. Scientific realism about computational approaches to the study of the mind can, however, be defended but only insofar as we treat the rule-governing aspect of computationalism as an idealization. So, it is possible to talk about nervous systems in computational terms. However, one should not take this kind of talk literally. Similarly, it is possible to talk about nervous systems as representational-using systems. However, one should not take discourse involving neural representations literally.

Explanatory unification and idealization in cognitive science Explanatory unification, or the urge to 'explain much by little', serves as an ideal of scientific modeling throughout the sciences (Mäki 2001). There are at least two senses of "explanatory unification." In one sense, explanatory unification is what we have been dealing with so far: the attempt to build simple, even if idealized, models of target systems. In a different sense of the term, it refers to the representational ideal of arriving at a single, unifying theory of some target domain. For example, the RTM and its different versions all aspire toward explanatory unification in the second sense of explanatory unification. So it is with the CTC and its different versions.

Explanatory unification is, however, in prima facie tension with a pervasive modeling technique in science. This is the practice that Weisberg (2007, 2013) refers to as *multiple-models idealization* (MMI). MMI implies that the search for a single, unifying model of a target system is likely a search with a dead end. If this is correct, then the aim of formulating a unifying theory of computation is likely to be a misguided goal. In the philosophy of cognitive science, MMI is now being taken up by researchers seeking to explain the workings of the mind by using multiple different and highly idealized models. As Sprevak and Colombo put it in *The Routledge Handbook of the Computational Mind*:

> The plurality of approaches within the computational sciences has helped to motivate several epistemological and methodological views going under the general banner of "pluralism." According to these views, the plurality of computational approaches we observe in the sciences is an essential feature of scientific inquiry into the mind. (Sprevak and Colombo 2018, 3)

In chapter 8, I aim to defend the plausibility of explanatory unification. I shall do so by framing the discussion with a distinct problem discussed in the philosophy of science. This is the *problem of inconsistent models* (PIM). PIM asserts that scientific modeling often consists of using multiple different models with inconsistent properties to model one and the same target system. If correct, PIM is a threat to scientific realism. It is therefore also a problem for a defense of scientific realism in cognitive science. I argue that PIM rests on literalism about scientific discourse. Once literalism about scientific discourse is problematized, PIM is no longer a thorn in the side of scientific realism. Building on chapters 3–4, I argue that one can start with MMI (minus PIM) and proceed toward explanatory unification in cognitive science.

Grand unifying theories of cognition: A case study In chapter 9, my aim is to bring all the many and varied threads discussed so far together under one grand unifying theory of the brain and cognition. A *grand unifying theory* (GUT) is an extreme form of explanatory unification. A GUT plays the same role as the "One Ring" in the *Lord of the Rings* universe, although nothing bad needs to happen. I consider the free energy principle (FEP) in theoretical neuroscience as an example of a GUT in chapter 9. Specifically, I argue for the following claims about the FEP:

- The FEP is an idealized, abstract, and mathematical model (chapter 1).
- The FEP is not literally true of its intended target domain (chapters 2–4).
- The FEP is consistent with a version of scientific realism based on the nonliteral view of idealization (chapters 3–4).
- The FEP does not lend support to the RTM (chapter 5).
- The FEP is a computational model yet lends no support to the CTC (chapters 6–7).
- The FEP represents an extreme form of explanatory unification (chapters 8–9).

The FEP serves as an ideal case study for the conclusions of this book. It can be used as a targetless model, a generalized model, and a target-directed model (chapter 1; see also Kirchhoff, Kiverstein, and Robertson forthcoming). It is a mix of a mathematical model and a computational model (chapters 1–3). It serves as an illustration of why it is a fallacy to confuse the mathematical specification of a computational model with the properties of a target system (chapters 6–7). Its mathematical basis lends no support to the philosophically motivated thesis with representationalism (chapter 5). It can be defended as a version of scientific realism but only once its idealizations are given a nonliteral treatment (chapters 3–4). Furthermore, the FEP is currently the subject of a heated discussion on the map problem (chapter 1). Recently, several authors have argued that because the FEP invokes formalisms from physics, machine learning, and probability theory to help understanding of the structure and function of the brain, it is best seen as a version of instrumentalism. Based on the nonliteral view of idealization developed across chapters 3–4, I argue that this understanding of the FEP is misguided. The FEP, I submit, provides a perfect case study in which to unify the themes discussed in this book.

5 Idealization and Neural Representation

Overview: A guiding principle of most cognitive science, including its philosophy, is that cognition is based on computational processes that manipulate internal vehicles with representational content. A common way of describing mental representations is in terms of mental states with *propositional content* (e.g., beliefs, desires, or intentions) and related propositional attitudes such as concepts, reasoning, and understanding. Three issues are worth highlighting here. (1) Mental states are usually taken have intentionality, in the sense that they are about states of affairs in the world—for example, I believe that there is milk in the fridge. (2) Mental states are usually understood to play a causal role in the generation of behavior—for example, my belief that there is milk in the fridge and my desire to have some milk may cause me to go and grab some milk. (3) There is a sense in which our respective mental states qua internal causes of behavior are "hidden" from others—for example, you cannot directly "see" my belief that there is milk in the fridge (you may infer that I believe such and such by observing my behavior). This view that cognition involves the manipulation of representational content is known mainly as the representational theory of mind (RTM).

I formulated (1)–(3) as a folk psychological version of the RTM. I shall not pay any attention to this version of the RTM. My interest is in neural representation, since the discussion about neural representation shares important connections with the discussion about scientific models that are not so obviously shared with the folk psychological formulation of the RTM. Nevertheless, realist defenses of neural representation are committed to the general idea that: (1) physical (neuronal) vehicles carry representational content; (2) the vehicles and/or representational content are causally efficacious in generating outputs, of various kinds; and (3) neural representations are hidden from direct observation but can be seen indirectly via the use of neuroimaging techniques.

In this chapter, I argue that appeals to neural representation is ultimately appeals to idealized constructs in cognitive science. The position in the literature that comes closest to this view is mental fictionalism. Mental fictionalism treats discourse about mental representations as useful fictions, where representational talk is literally false but useful for explanatory purposes (see, e.g., Toon 2023). Fictionalism about mental or neural representation is a way of treating such theoretical posits as idealized posits are treated in model-based cognitive science. What follows? We can make use of neural representational posits, just as we can make use of point objects in physics. However, as with point objects, one should not infer that nervous systems are literally representational systems. I agree with the mental fictionalist on this point. However, unlike concepts such as point objects, ideal gasses, or infinite populations, it is much less obvious that discourse about mental or neural representation facilitates genuine explanation. The reason is that making use of such theoretical terms is usually taken to imply something true about the brain and cognition. This much is off the table if talk about mental or neural representations are ways of idealizing the brain and cognition. Moreover, if such talk is ultimately untrue, then insisting that it is nevertheless useful might be theoretically misleading.

This is what I aim to establish. In doing so, I set aside any further discussion of mental fictionalism until chapter 10. In this chapter my focus will be on realism about mental or neural representation. I argue that it is theoretically preferable to treat discourse about mental or neural representation as discourse about idealization in cognitive science.

5.1 Introduction

Discussing mental representations as idealizations is uncommon. However, idealized treatments of mental representations are easy to find. Take, for instance, the tradition, following on from Dretske (1981), seeking to provide a naturalized theory of representation. In Dretske's theory of representation, *nomic covariation*, states are assigned content via the following kind of definition: a state S_b represents *Bs* as *Bs* if and only if the system tokens S_b when and only when *Bs* are present. For example, S_b represents frogs as frogs if and only if the system tokens S_b when and only when instances of froghood are present. The most widely discussed problem for covariation theories of this kind is called the *disjunction problem*. This problem arises because of the "when clause" of the definition of nomic covariation. According to this clause, S_b appears to represent froghood. However, say the visual conditions are suboptimal and that one day the system tokens S_b when a snail is present. The "only when clause" of the nomic covariational definition of representation allegedly requires that the system does not represent froghood simpliciter. Instead, the system in question represents the disjunction "frog or snail." Hence, when tokened, S_b represents "either frog or snail." To prevent this outcome, nomic covariationists have attempted to defend their view by appealing to *ideal perceptual conditions*. Wallis defines the idealized nomic covariational position as follows: "A state S_b represents *Bs* as *Bs* iff *under ideal circumstances* the system tokens a S_b when, only when, and because *Bs* are present" (1994, 409). The disjunction problem goes away but only because one idealizes error away from the definition of covariational approaches to representation. Idealizing error away from the definition of representation, however, comes with a significant cost: one ends up idealizing away one of the core properties of representation: misrepresentation.

Following in the footsteps of covariational approaches to representation is the influential framework of *teleosemantics*. Like the nomic covariational framework, the teleosemantic program seeks to give a naturalistic explanation of the semantic properties of representations (Godfrey-Smith 2006). Teleosemantics seeks to show how internal states of an agent could, as a matter of mind-independent fact, represent aspects of the world by appealing to the concept of biological function (Millikan 1984). At the core of Millikan's teleosemantics is the idea that all representations are signals sent from a *producer* (some external cause) to one or more *consumers* (internal states of an organism). Hence, for Millikan, representations stand between producers and consumers (Schulte 2023). The notion of a "consumer" has a prominent place in teleosemantics. Briefly, teleosemantics seeks to give etiological functions a content-determining role and is explicitly committed to the importance of a representation consumer, where a "representational consumer" is an identifiable system that can take a representation as input and, on the basis of doing so, is able to generate an output in response (see, e.g., Millikan 1984). Yet, the theoretical posits of a representational consumer is taken by some to be an idealization. For example, Shea—a realist about mental representations—says that there "need be no identifiable subsystem that takes this collection of representations as input and produces behavior as output" (2020, 19). Shea also notes that:

> It is very hard to see a principled way to identify representation consumers in the brain, if consumers are devices whose output fixes content. Even *idealized* diagrams of neural circuitry are notoriously interactive, consisting of a complex mix of feedforward, feedback, and lateral connections, some proceeding in a series of levels or layers, others cross-cutting or bypassing those layers. (Shea 2020, 19; emphasis added)

Coelho Mollo draws similar conclusions:

> The separability of representations and their consumers is an *idealisation*: It ignores the close interdependencies between most parts of the system, simplifying its organisation so as to bring to the fore a useful distinction between different functional roles; distinction that does not exist in such a neat way in the actual system, but that makes it more comprehensible and tractable. (Coelho Mollo 2021, 1058; emphasis added)

The idea of a representational consumer relies on a set of modeling assumptions—for example, that one can decompose complex systems into discrete parts with localized functions. Yet, as we shall see in chapter 7, these kinds of modeling practices are ultimately a way of imposing a set of theoretic assumptions on the nature of the system under study—and crucially a set of assumptions that the target system being investigated need not itself manifest. So, discourse about representational consumers can possibly be seen as a way of simplifying complexities in target systems, even if one should not impute such features to the actual target system under examination. Now, if the target system in question does not literally have consumers, then it is hard to see how an idealized treatment of said target system will prove explanatorily useful.

Millikan (2000) is, perhaps surprisingly, very clear about the idealized status of the teleosemantic framework for naturalizing representational content. Millikan states:

> What you must say, says Tilly, is that the representations represent what would be in the world, running isomorphic to them, if the cognitive systems were operating correctly. That is, what a teleological theory of content does is to take some more basic theory of content, point out that the application of that theory to actual creatures *requires idealizing them in certain ways*, and then *offer the teleological principle to explain which idealization is the right one to use in interpreting intentional contents*, namely, the one that fits how the cognitive systems were designed or selected for operating. (Millikan 2000, 229; emphasis added)

Or consider what Godfrey-Smith says of the general project of naturalizing mental representations:

> The naturalistic philosophical literature has spent *a lot of time describing idealized hypothetical scenarios* that have reasonable or compelling descriptions in both physicalist and semantic terms. The aim has been to describe as minimal and empirically feasible a structure as possible, while maximizing the number of features of paradigmatic semantic description that attach plausibly to the structure. (Godfrey-Smith 2006, 60; emphasis added)

It is widely known that explaining cognition by appeal to neural representations raises some fundamental questions, and related problems. Among the most difficult are problems pertaining to naturalizing the putative semantic content of mental representations—their aboutness and truth-conditionality. What makes this a particularly difficult project to succeed with is that naturalizing content is taken by many to require a satisfying response to the following cluster of problems:

- *The source of original content*: The color "red" at a traffic light has the meaning it has *derivatively*, in the sense of being grounded in sociocultural norms and conventions. By contrast, if cognitive systems are cognitive partly in virtue of having internal states with semantic content, the content of those representations must be *nonderivative* (i.e., original).
- *The causal efficacy of semantic content*: Having causal powers is often said to be evidence of being real. Therefore, if neural states have content, then the content of those states must have causal powers. In other words, the content and not merely the physical vehicles of internal representations must be causally efficacious.
- *The determinacy of content*:. When a "frog" flicks its tongue, it does so because there is a "fly" in its visual field. It flicks its tongue because the content of its internal representations is *determinate*—that is, the content is such that it is either true or false. What we do not want of a theory of mental representation is that the content posited is nondeterminate—for example, the frog flicks its tongue and its content is "there is a fly or there is a black dot."
- *The possibility of misrepresentation*: Insofar a system has representational capacities, it must also be possible for that system to misrepresent. That is, error must be possible.

These issues have no agreed solution in the literature. As Piccinini notes, the "difficulty of one or more of these problems has led many theorists . . . to reject neural computations, neural representations, or both" (2022, 3).

Coelho Mollo (2021) has recently made an intriguing alternative suggestion, starting from the idea that representational models of the mind embody idealizations and, therefore, a degree of mismatch between their posits and the actual properties of cognitive systems is inevitable. Coelho Mollo (2021) argues that it is because our best models of neural representations are partial and idealized that these kinds of problems cannot be solved. Or, differently put, we imagine that neural representations have determinate content that plays a causally efficacious role, and so on, when in fact the nervous system does not trade in states with those kinds of properties. Coelho Mollo concludes:

> Indeterminacy of content is the result of the limitations of our models and of our knowledge; limitations that are unlikely to ever go away, given the complexity of the systems modelled, the epistemic opacity of the relevant factors, and the consequent partiality of our models. That indeterminacy of content marks all our attempts at giving a realist theory of representation is not the expression of a metaphysical problem, but rather a consequence of the adoption of an idealised model to explain certain phenomena. (Coelho Mollo 2021, 1060)

Consider the much-discussed distinction between representational vehicle and representational content. Ramsey (2023) asks: "Why do we call representational entities '*vehicles*' and their relation to things they represent as having '*content*'?" Ramsey's assessment is as follows: "This implies that the representational entities [the vehicles] are serving something like mere *containers* that, as such, have the job of holding or carrying their more significant cargo. How did this *strange terminology* come about, this language suggesting that the intentional relation is somehow encased in the representational entity?" (2023, 4; emphasis added). According to Ramsey:

> While the details of this picture needn't concern us, it is worth reflecting on this conception of mental representation and what it implies. It is a conception that treats the representational vehicles as *vessels for conveying freight.* It regards representational contents as *important and mysterious somethings*, whereas the representational entities themselves, qua-packaging, are regarded as having diminished significance. (Ramsey 2023, 4; emphasis added)

What this suggests, if correct, is that the very distinction between vehicles and content is an idealization. It is worth emphasizing here that just because X is an idealization, it does not follow that X is explanatorily useful. In other words, if there is no genuine distinction between vehicles and content, then the continued use of this distinction is theoretically confusing.

More evidence for our representational models of the mind being idealized models comes from the fact that cognitive science is a model-based science. This is nicely captured by Kriegeskorte and Douglas: "The goal of cognitive computational neuroscience is to explain rich measurements of neuronal activity and behavior in animals and humans by means of biologically plausible computational models that perform real-world cognitive tasks" (2018, 1148). A famous example of this kind of model building is the discovery of place cells in the rat hippocampus (O'Keefe and Burgess 1996). Place cells are individual neurons whose activity is specific to an animal's location in space. The target system is the navigational activity of rats and the contribution of place cells in the hippocampus to the success of this activity. In O'Keefe and Burgess (2005; see also Shea 2020), the model represents the firing rate plots for thirty-five place cells in the dorsal CA1, where different cells are interpreted as correlating with different spatial locations in a maze as the rat moves around. The model is used to explain the target system in the following way. Each of the thirty-five plots show the sensitivity of an individual cell, where the shading refers to the location of the rat when that cell is active. The darker the shading, the higher the firing rate. The array of neurons, and their respective firing rate, have been shown to provide an accurate measure of the rat's spatial location in a maze. Yet, the model is an idealized model (Orlandi 2020). For instance, O'Keefe and Burgess (2005) note that the place-cell "theta phase" functions to provide a periodic clocking system against which the timing of hippocampal spikes can be measured—for example, cell 1 fires when and only when the rat is in the top corner of the model, and so on. However, the assumption that place cells exhibit strong periodicity is an idealization, since many cells in the entorhinal cortex are far from having spatially periodic firing. Orlandi confirms this when stating:

> Place cells also exhibit remapping where the same place cell might selectively respond to a different place when the animal is exploring a different environment. *The example described here is indeed simplified and idealized,* but its main point is to show what it might be like for rats to use maplike structures. Whether place cells in rats also have a context-sensitive response behavior is less central. (Orlandi 2020, 124n10; emphasis added)

What we get is a hypothetical scenario; one that describes what it would look like if rats were to use such maplike structures to successfully perform certain navigational tasks.

Idealizations aside, many find this kind of example, or something like it, a compelling reason to think that neural representations are needed to explain cognition. The example above sketches a picture in which patterns of neural activity can be argued to encode information about spatial locations in the world, which can be exploited to navigate a maze. Shea says: "The 'cognitive map' in the rat hippocampus illustrates how use of structural correspondence to perform task functions can be the basis of representational content" (2020, 116). Cao agrees: "Place cells represent places and grid cells represent locations in space" (2022, 15). Crucially, what both Cao and Shea agree on is: embedding representational posits within a model of cognitive behavior makes the system under investigation tractable. Tractability, however, is one of the central functions of idealization in scientific modeling given the implausibility of creating hyper-realistic models of very complex target systems (from chapters 1, 3–4). Tractability is often associated with one of the functions of idealization. Point objects in physics facilitate computational tractability. Yet, they are not part of target systems in reality. Even if one grants that appeals to mental or neural representations facilitate tractability, it does not follow that such representational posits are part of actual target systems.

This chapter argues that even if it can be shown that there are such explanatory gains, the realist about neural representation faces a problem. They fall short of meeting what I call the *translation task*: the task of showing that it is literally true that the representational features of a cognitive model correspond to features of the target system of interest.

The structure of this chapter takes the following form. In section 5.2, I show how realist proposals about neural representations fail what I will call the "translation task." In section 5.3, I argue, on this basis, that neural representations are best understood as idealizations. My method will be to employ the nonliteral view of idealization (chapter 3) as my template of analysis. In section 5.4, I discuss three implications for understanding neural representations as idealizations: (1) discourse about neural representation is useful, cannot be eliminated, and should be treated nonliterally; (2) discourse about neural representation is useful, can be eliminated, and should be treated nonliterally; and (3) discourse about neural representation is useful, irrelevant for explanation, and should be treated nonliterally. All three implications are hostile to any kind of entity realism about the existence of neural representations. However, none of them threatens a realist view of scientific models that include representational constructs for the purposes of explanatory gain (of some kind). I conclude in section 5.5.

5.2 The Translation Task and the Representational Theory of Mind

I will exemplify the translation task via the tried and tested ideal gas law.

Say you are a medical gas provider for hospitals in need of transporting patients some distance. You need to figure out how much available oxygen is needed for a patient transfer lasting 42.5 hours. One option would be to simply make a guess, fill up some gas cylinders, and hope it all goes well. Failure would be costly and the risk high for everyone involved. A better way to proceed would be to construct a mathematical model based on the ideal gas law. Here is the crucial point: however carefully these models are created, and however diligent you are in fine-tuning their parameters, you are not studying these models for their own sake. You must be able to transfer the specified model parameters into claims about the target system, in this case, specific claims about precisely how much oxygen is needed to transfer a ventilated patient over a journey of 42.5 hours. This is what I call the *translation task*: turning a model of the world into truth-conditioned assertions about the world. In this case, the assertions had better be true, or accurate—or close enough to being true that the patient makes it from one hospital to the other. With the ideal gas law at hand, one can make the following computation: Take an E cylinder of medical oxygen with a physical volume of 4.7L and a pressure of 137 bar (equal to 13,700 kPa). If one applies the ideal gas law at room temperature, $P1 \cdot V1 = n1 \cdot R1 \cdot T1$ (inside the cylinder) and $P2 \cdot V2 = n2 \cdot R2 \cdot T2$ (outside of the cylinder), it results in the following: $P1 \cdot V1 = P2 \cdot V2$. These formalisms can be used to derive the result that a 4.7L cylinder with a pressure of 137 bar is equal to 637 liters of oxygen. This allows for a further calculation of having 42.5 hours of available oxygen in a full tank (Candan and Cascella 2022). Even though the ideal gas law is an idealization, it allows one to turn the model into truth-conditioned assertions about the world. In this way, the ideal gas law illustrates one way of addressing the translation task. It facilitates solving a particular mathematical (and practical) problem.

Things are less straightforward in the discussion about neural representation. One key reason is that idealized posits in science and idealizations in cognitive science appear disanalogous. Most cases of successful idealization in science involve particular systems, such as natural gases, where the scientists working with some model, such as the ideal gas model, have a very solid understanding of the parts involved in the model and their respective functions and properties. This is what allows scientists using the ideal gas model to deliberately treat the value of intermolecular interactions as zero. As Wallis puts it:

> The molecular theory of temperature provides one with a model for understanding pressure and temperature in a gas. Given the molecular model of temperature, one can, in many cases, idealize away from negligible forces and from the inelastic component of collisions so that one can formulate an idealized law for that portion of the domain. (Wallis 1994, 425)

There is, however, no such understanding about neural representations. Some argue that neural representations can be observed (Piccinini and Thompson 2018). Others deny that neural representations can be observed based on neuroimaging (Facchin 2024). Some argue that neural representations exist (Shea 2020). Others deny that neural representations exist (Hutto and Myin 2013). Some think that the existence of neural representations requires the presence of representation consumers (Millikan 2000). Others deny this (Coelho Mollo 2021). Some argue that the content of representations must have causal efficacy (Piccinini 2022). Others deny this and think that all the causal efficacy of representations involve the vehicles rather than the contents themselves (Egan 2020). The list of disagreement goes on.

Another reason for the disanalogy is that idealization in science is recognized to have distinct functions. They flag that certain variables are causally irrelevant to explanation (this is the case in the ideal gas model); they sharpen our focus on specific causal factors rather than others (this is also the case in the ideal gas model); and they facilitate computational tractability (which is the case in the baseball example discussed in chapter 3). What about mental states? Or, what about appeals to internal representational content? Is such a treatment explanatorily useful? In the case of the ideal gas model, idealizations set aside what Strevens (2007) calls causal nondifference-makers. That is, by setting the value of intermolecular interaction to zero scientists acknowledge that this value can be neglected in order to focus attention on the actual causal difference makers (i.e., pressure, volume, and the number of moles in a gas). This is not true of idealized posits such as mental states and neural representations. Treating mental or neural representations as idealizations does not flag that appeals to mental or neural representations are explanatorily irrelevant. Moreover, treating mental or neural representations as idealizations does not sharpen our focus on the actual causal difference-makers underlying the behavior of explanatory interest. Finally, appeals to mental or neural representations as idealizations do not solve any intractable problems; nor does such an idealized treatment of mental representation facilitate computational tractability. Ultimately, the real value of idealization in cognitive science is that it gives us reason to treat talk of an inner world of contentful mental representations as merely metaphorical - as a means of acknowledging that we should not, and could not, mean what we are saying when making use of these theoretical terms (Sganzerla, Hutto, and Kirchhoff, forthcoming).

With these points in mind, I now turn to discuss three realist proposals about neural representation. I consider Cao (2022), Shea (2020), and Shagrir (2012), using their accounts as representatives for representational explanation in the philosophy of cognitive science.

5.2.1 Cao's Account of Representation

Neuroscientists are interested in neural representations. A neural representation is a "functional pattern of neural activity" (Cao 2022, 2), and is best understood via an exploration of how "the physical internals of organisms are arranged, and how those arrangements are connected to the external world" (Cao 2022, 3). As Cao is quick to point out: "Because these facts are so numerous and so complex, we need some way to make them more *cognitively tractable*" (2022, 3; emphasis added). In chapter 3, I mentioned that one of the central functions of idealization is that it enables researchers to solve certain analytically or computationally intractable problems. Cognitive tractability is not the same as computational tractability. Yet, it serves a similar kind of outcome: understandability. One possible implication is: neural representational discourse may be used to make sense of the dynamics underlying behavior even if the target system itself is nonrepresentational. Such posits enable the construction of simpler and idealized models of a target system. We can get a sense of this by considering this quote from Chirimuuta:

> Critics and doubters of neural representations often argue that less demanding notions of correlation or triggering would be just as adequate to describe sensory neurons' reliable responses to specific distal stimuli, with less of the contestable baggage that comes with the positing of neural activity as representing something. However, we can now see that just invoking the less demanding notion of correlation between neural activity and extradermal state does not facilitate the exclusive focus on the distal object needed by the cognitive neuroscientist. This is simply because *there are too many processes, which can equally be said to correlate with the brain response, along the mediating chain from the distal object.* Similarly, the notion of triggering does not isolate the distal object. Triggering is a causal concept which means that it opens the door to holding that any item or event along the mediating chain is a trigger of the neural response that it causes. In short, *the proposal here is that appeal to neural representation allows an important simplification in cognitive neuroscience because it facilitates and justifies the black-boxing of intermediaries between brain responses and the distal objects that are relevant to the explanation of behavior.* (Chirimuuta 2024, 160; emphasis added)

Abstraction is the practice of omitting certain features of a target system, usually to make other features more salient—for example, for explanatory purposes (chapter 2). Once one must use models that are more cognitively tractable than the target system, one is inevitably in the game of constructing simplified, and often idealized, models of target systems. The problem for the realist is that this is precisely how idealizations such as point objects or infinite populations are used in physics and biology. They simplify and sharpen explanatory focus of certain aspects of target systems. But they do not exist on the side of the target systems.

How does the appeal to neural representations fit in? Cao says:

> A . . . way of arriving at representation-talk may come from our trying to make sense of these innumerable and complex wiring-and-connection facts. Perhaps finding real patterns in the dependence of internal activities on how the world is (and vice versa) allows us to compress otherwise unwieldy bodies of information. If so—and perhaps only so—we would have good reason to claim that there are representations in the brain. (Cao 2022, 3)

This view is initially very attractive. One way to put it would be that it explores how patterns of neural activity are systematically correlated with an expected change in behavior, allowing one to model the target system as if it has a representational function. Cao (2022) does not put it quite like this. Instead, Cao thinks that the appeal to neural representations serves a particular explanatory function, one that creates an explanatory bridge between the wiring-and-connection facts about the nervous system and the interpretation facts (I take this to be reasonably like the point made by Chirimuuta [2024] above). The so-called interpretation facts are said to be "manifested in our practices of interpretation and ascription of content" (Cao 2022, 3). In this sense, the interpretation facts are explananda and, therefore, not an independent way of explaining neural representations. However, Cao notes:

> What neuroscientists are more interested in, by contrast, are not the "interpretation facts", but rather the "wiring-and-connection facts"—that is, how the physical internals of organisms are arranged, and how those arrangements are connected to the external world and interesting behaviors exhibited by those organisms in that world. (Cao 2022, 3)

This way of putting things allows for the following kind of attitude toward the function of neural representation: neural representations do not strictly speaking exist on the side of the target system under investigation. The claim is not that this implies any kind of eliminativism. Instead, the claim is that even if neural representations do not exist on the side of the target system, they exist on the side of the model used to explain the target system. Hence, one option is that the appeal to neural representation is a useful and legitimate practice, even if representational posits do not exist in the target system. The interpretation facts are the idealizations. Cao (2022) comes close to this conclusion in her discussion of capacities that in the literature have been deemed "representation-hungry" (Clark and Toribio 1994), such as memory, planning, imagination, and so on. As Cao notes:

> The relationship between complex behavior and internal representations is so tight that it is not possible (and perhaps not even conceivable) for a creature to display complex behaviors and dispositions without it having representations. So once again, it seems as if representations are explananda—another (counterfactually robust) way of describing the behaviors to be explained, but not an independent resource for explaining them. They are just another way of describing the "interpretation facts" (Godfrey-Smith, 2004), which are manifested in our "practices of interpretation and ascription of content." (Cao 2022, 3)

A different issue concerns the representational description used to explain the functional capacities of neural activity. A familiar view is that representation depends on representation "users." Cao (2022) considers three kinds of users: (1) downstream consumers of representations within an organism, (2) the whole organism or system, and (3) the scientists and their use of representation. I discussed (1) above, and identified it as an idealization. Cao focuses on (3)—the scientists and their use of representation. The justification is: "The scientist posits representations to explain how the system works. So the representations had better be involved and in some sense causally efficacious in how the system works" (2022, 15).

This conclusion does not follow. That is, the claim that if scientists posit representations to explain cognitive capacities, then representations must be doing some causally relevant work does not follow. To see this, let us return to the ideal gas law. The law assumes that one can model long-range gas molecules as noninteracting (an idealization). This assumption is false, if taken literally (following from chapters 2–3). It is an idealization, where the idealized posits do not correspond to anything in the world. Yet the idealization is taken by most to have epistemic value. For example, it allows one to interconvert between volumes and molar amounts in chemical equations—when having to determine the nitrogen density in a hot air balloon, for instance. So, a scientist can posit an idealization to explain some target system without that construct being causally relevant for how the system works.

The appeal to neural representations can serve the same function. According to Cao, the function of neural representation in scientific modeling is that it "allows us to compress otherwise unwieldy bodies of information," into tractable claims about the target system under investigation (2022, 3; see Chirimuuta [2024] above). The function of neural representation is, therefore, to remove us from the messy details of the brain. It is another clear-cut example that discourse about neural representation serves as a way of simplifying a model for the purposes of cognitive tractability. In other words, the function of neural representation—if it is to serve to make the target system more tractable—is to present information about the nervous system in a way that is simple enough for us to deal with. Here is the problem for the realist: if what is really going on is things to do with the wiring-and-connection facts, then adding a layer of discourse about neural representations is not the same as saying something about the wiring-and-connection facts. One can take such talk seriously. However, if neural representation is an idealization, then is not true of the brain and cognition.

There is more to be said. According to Cao, what neuroscientists refer to as representations "does not require any normativity" (2022, 7), but only that patterns of neural activity can be intervened on in such a way as to produce certain kinds of causally reliable outcomes. Cao says: "Not much would be lost if we understood the notion of representation in neuroscience in a 'merely' causal way" (2022, 7), where neural responses to stimuli "are *presumed to be* representations partly because they are thought to be levers for intervening" (2022, 5; emphasis added), and nothing more than "selective responsiveness to certain classes of stimuli" (2022, 7). The problem with this account should be familiar. The capacity of patterns of neural activity to produce certain kinds of reliable outcomes, even if this is their proper function, falls short of establishing that those patterns have truth-conditions (Hutto and Myin 2013). Standard realist views on representation require that for *X* to be a representation, *X* must have truth-evaluable content. However, this condition cannot be met by the proposal that representation in neuroscience is best understood in a "merely" causal and correlational way. There are at least two ways of reading the implications of this for Cao's (2022) account of neural representation. One is that neural representation is just the wiring-and-connection facts that neuroscientists are interested in. That is fine. However, then labeling these wiring-and-connection facts as representational is just that—labeling. This point can be further motivated by appeal to Ramsey's (2007) fifth adequacy condition on a theory of representation. A theory of representation should not overexplain representational capacities in such a way that it turns out to dispense with representation: for example, if it turns out that representations function as "mere causal relays" then, in effect, the phenomenon of interest—the contentful, truth-conditional state—has disappeared. A different way to assess the situation would be to say that insofar as Cao (2022) is correct in her assessment, then positing neural representations to explain cognitive activity is to make use of an idealization: it is false of its target system, if taken literally. Since this is where Cao's treatment of neural representations ends up, to insist on the explanatory relevance of representation is to insist that something untrue of the target system is relevant for explanation of cognitive activity. There is no problem with this in principle—it requires only acknowledging the idealized status of neural representations in scientific modeling.

5.2.2 Shea's Account of Representation

Shea (2020) is another prominent defender of neural representations. For Shea, realism about neural representations is a commitment to there being objective, mind-independent vehicles of content; that is, individuated physical particulars that function as bearers of semantic content, whose causal interactions explain behavior. Let us start by considering some specific examples mentioned by Shea.

- Case 1. A primate is trained to choose between two sets of objects. It is rewarded if it selects the set with a higher number of items. The basic idea is that if presented with two buckets containing collections of objects, to pick the bucket with more objects is to perform the task successfully. Shea states: "The training has used and tuned the animal's analogue magnitude subsystem, giving the animal a disposition to pick the thing in the world corresponding to whichever analogue magnitude register in its parietal cortex is more active" (2020, 199).
- Case 2. The firing mechanism of a rifle can be described as consisting of five linear and factorized steps: (1) the user's finger moves backwards; (2) the trigger moves backwards; (3) the firing pin shoots forwards; (4) the charge in the cartridge ignites; and (5) the bullet flies off at speed.
- Case 3. A magnetotactic bacterium moves in the direction of oxygen-free water. As Shea puts it: "As the case is standardly described . . . , moving in the direction of oxygen-free water is a stabilized function of the bacterium's behaviour, but not a robust outcome function" (2020, 203).

For each of these three cases one can ask, as Shea (2020) does: Do we not have within reach a nonsemantic, or noncontentful, causal explanation for how behavioral output is generated for each of these cases? Realists about neural representation are committed to there being such a nonsemantic causal description. This follows from their commitment to explaining representational content in noncontentful terms, as part of the project of naturalizing representational content. The question, therefore, is: If such a noncontentful description can be given, what do we gain from adding representational content to the mix? Shea says: "Representationalists also typically subscribe to the claim that representations interact causally in virtue of non-semantic properties. If so, a complete causal description of the behaviour of the system is available that doesn't mention semantic properties at all" (2013, 498).

If this is possible, then the realist does not have a straightforward way of successfully handling the translation task. Here is how Shea responds: a nonrepresentational alternative would have to "replace content-based explanation with much more fine-grained explanations. The trouble with these views is their complexity. More complex properties are generally less good candidates for explanation" (2020, 205). Chirimuuta (2024) draws the same conclusion. Does this follow? Let us start by making the following observation. Ever since scientific modeling has attracted the attention of philosophers, they have emphasized how models simplify target systems for ease of explanation. So, simplification is a technique used to make explanation tractable in science given its often very complex subject matter. This issue turns on cognitive tractability, discussed above. Hence, one option here is to say that it is the representational structure of a model that makes it a simple model of its target system. This might very well be the case. What follows? Take the notion of a point particle in physics. A point particle, or ideal particle, is an idealization. It is a construct often invoked in physics. The defining feature of a point particle is that it is dimensionless—it lacks spatial extension (see chapter 3 for a specific model). According to the standard view of idealization (chapter 2), most of philosophy of science treats idealizations as the process of assuming facts about some target system being modeled that are strictly speaking false but that make the model in question easier to understand or the problem in question easier to solve. Content-based explanations of cognition may play this kind of role—they may help us suppress huge amounts of information in the service of cognitive tractability, even if content-based explanations are idealizations in the same sense that point particles are idealizations. Of course, this does not give the realists what they need to secure realism about neural representation.

Let us now look to cases 1–3 for a possible way of handling the translation task for a defender of realism about neural representation. According to Shea (2020), one does not gain anything by including representational content in one's explanation of cases 2 and 3. However, for cases such as 1, one does gain something in understanding via representational explanation. Let us focus on the difference between cases 1 and 2. It is possible to provide an explanation of case 2 in terms of representational content. For example, if the firing pin carries representational content, "movement of the user's finger leads to a representation being tokened with the content *the user's finger has been pulled back, fire a bullet,* which leads to a bullet being fired" (Shea 2020, 201). Yet, this representational redescription marches in lock-step with the factorized nonrepresentational, causal explanation. Nothing is gained other than mere relabeling. Here is the apparent difference maker for taking case 1, and not case 2, as an example where there is a need for representational explanation. This is the model invoked to explain the target behavior in case 1 (figure 5.1).

The basic idea is that the model captures aspects of the target system because of the following considerations: on the input side, analog magnitude registers in the parietal lobe correlate with the numerosity of collections of objects in the distal environment, which is a correlation mediated by a variety of different proximal inputs; on the output side, the primate's reaching for the more numerous bucket of items is mediated by a variety of different motor outputs. Hence, "there is bridging at input and at output" (Shea 2020, 203).

Figure 5.1 is an idealization. It assumes input to be discrete. It depicts bridging states as discrete, as well as the output states. Furthermore, figure 5.1 provides an example of idealization in the relevant sense:

> Idealizations . . . serve the function of *isolating a causal mechanism* or a simple dependency relation. Many models in natural and social sciences are *minimal models* in this sense: they help focus on the structure and operation of one mechanism at a time, in isolation from all other mechanisms and conditions that in actual circumstances may also shape the behavior of the phenomena under investigation. (Mäki 2020, 219)

What makes this a positive case for realism about neural representation? Shea's (2020, 2013) basic view on this question appears to be: embed a

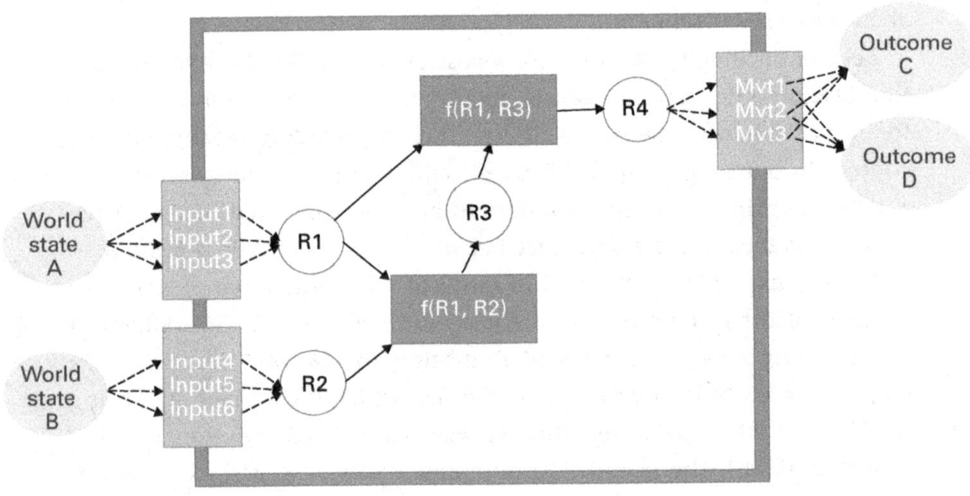

Figure 5.1
A model description of bridging at input and output (Shea 2020, 202).

system in some environment, and entities within the system acquire a new set of relational properties. Shea explains: "The 'information' of information theory is a matter of covariation, so embedding a system in an environment gives rise to a host of new informational properties" (2013, 498).

What new informational properties? One could certainly say that a system acquires the ability to respond to a large distribution of environmental conditions. Yet, this is the same as saying that such a system is sensitive to a large array of correlations. An immediate difficulty with this way of seeking to justify realism about neural representation is its starting point: *covariation*. Covariation expresses a quantifiable relationship between the values of different variables such that if *A* tends to increase (or decrease) in value, the corresponding values of *B* tend to also increase (or decrease). There is a robust, corresponding relationship between *A* and *B*, if *B* increases when *A* increases. Or, there is a robust, corresponding relationship between certain kinds of inputs and outputs conditioned on mediating structures. This is what is illustrated in figure 5.1. The only thing that is added is that there may be more than just one pair of input-output mappings. Figure 5.1, therefore, represents systematic correlations between certain kinds of inputs and certain kinds of outputs.

We have good reasons to think that semantic content cannot be derived from information in the sense of covariation (Hutto and Myin 2013). In other words, if information is covariance, then information is not any kind of semantic content—that is, information as covariance does not possess the special properties of semantic content, such as reference, implication, and truth. Shea (2013) is right that, once embedded in an environment, certain systems are able to make use of covariance in order to gain an adaptive advantage—for example, systematically related variables can be exploited in order to minimize entropy (for discussion, see chapter 9). But this still does not get us to reference, truth, and implication.

Starting with covariation makes it impossible to address the translation task, since (1) covariation is everywhere in nature—and, therefore, does not cut in cognitive joints—and (2) covariation does not entail any kind of semantic content. I am not saying that Shea (2020) thinks that covariation is sufficient for representation. Rather, on Shea's account, covariation is one of the central ingredients that becomes available to a system when it is embedded in an environment—other conditions must also be met for genuine representation, such as normativity from robustness, learning, selection, and so on. Yet, consider the "covariation doesn't constitute content" principle of Hutto and Myin (2013). The basic point of this principle is that "anything that deserves to be called content has special properties—e.g., truth, reference, implication—that make it logically distinct from, and not reducible to, mere covariance relations holding between states of affairs" (2013, 67). Hence, if information is nothing but covariation, it is not any kind of reference-making, truth-bearing content. To suggest otherwise would be to presuppose what needs to be explained. One can describe information as covariance as having semantic content. But this would be to imbue covariance with properties it does not have.

5.2.3 Shagrir's Account of Representation

A third type of representational account is Shagrir's (2012) structural or isomorphism view of representation (see also Shea 2020). Generally speaking, the idea behind the term "structural representation" is that representations preserve a formal or mathematical structure with the target domain that they represent. Or, in model-building terms, that the model and the target system share a structurally similar formal or mathematical structure. Structural representation is, therefore, a view of the model-world relation that can encompass "representational systems as family trees, maps, and various measurement schemes" (Shagrir 2012, 521).

According to Lee and Calder, contemporary "debates over the role and value of representation in cognitive science revolve around the notion of 'structural representation', or 'S-representation' for short" (2023, 1). Facchin notes that "neural structural representations . . . are absolutely central to the cognitive neuroscience revolution" (2024, 1). Philosophical work on S-representation conforms with current practices in the field of computational cognitive science. There is now a large literature on structural representation as internal models in the philosophical literature (see, e.g., Bartels 2006; Cummins 1996; Gładziejewski 2015, 2016; Grush 2004; Isaac 2012; O'Brien and Opie 2004; Ramsey 2007; Rescorla 2009).

The theory of neural representation presented by S-representation can be described through a set of four conditions that together are taken to support the view that the theory is explanatorily robust (see, e.g., Lee and Calder 2023; and also Gładziejewski 2015; Piccinini 2022; and Shea 2020). The criteria are: structural correspondence, action guidance, decoupleability, and system-detectable error or misrepresentation. Facchin (2024) presents these conditions as follows:

Within a system **S**, a vehicle **V** is the vehicle of a structural representation of a target **T** if and only if:

1. **Structural similarity: V** is structurally similar to **T**; and
2. **Action Guidance:** The structural similarity in **(1)** allows **V** to guide **S's** action in regards to **T**; and
3. **Decouplability:** (2) can obtain even when **V** is decoupled from **T**; and
4. **Error Detection: S** can detect the representational errors of **V**.

I will now sketch these criteria, and do so by considering the example of the function of place cells in the rat hippocampus, which I introduced earlier.

The *S*-representational account explains it as follows. First, the relation between the structure of the hippocampus and the task environment in one of structural correspondence, where some aspect of the relational organization of the task environment is correlated with the organization of the rat hippocampus. Second, the structural correspondence between the organization of the hippocampus and the task environment can be exploited for action guidance, where the firing rates of neural patterns of activity function as a navigational aid. Third, according to Lee and Calder, *S*-representations "are potentially *decoupleable* from that which they correspond to in guiding action" (2023, 4), inviting the view that rats might be able to plan their route ahead of time. Finally, whenever some unsuccessful behavior occurs—that is, whenever there is misrepresentation—the expectation is that there are mechanisms in place enabling the organism to update its internal configuration, which, in turn, increases the degree of correspondence between the structure and activity of the hippocampus and the task environment. Taken together, this is how the *S*-representational account can be applied to the discovery of place cells and their role in rat navigation.

Shagrir (2012) builds his account of structural representation based on Swoyer's (1991) analysis of the notion. For Swoyer, if a model and its target domain have "the same structural or formal features" (1991, 457), the two share a higher-order structure called a *shared structure*. If the model and the target system being represented by the model have a shared structure, then the model represents the target system in virtue of this second-order isomorphism.

This is precisely how similarity or structural correspondence is articulated in the literature on scientific modeling. Consider what Elliott-Graves and Weisberg say about the Schelling model (chapter 1):

> we think that the best way to describe the relationship between the Schelling model and Philadelphia, insofar as there is one, is as a *relationship of similarity*. Or more precisely, the Schelling model is *similar to a representation* of the City of Philadelphia target system in *certain respects and degrees*. If the model is similar enough, in the important respects, to the City of Philadelphia, then it will be considered a good model for explaining segregation. (Elliott-Graves and Weisberg 2014, 181–82; emphasis added)

Weisberg (2012) develops his similarity account from Tversky's (1977) "contrast" view of what makes two objects similar to one another. Tversky's account starts with the idea that the similarity of objects a and b depends on the features they share and the features they do not share. On Tversky's account, to assess the similarity between a and b, one needs to start by articulating some set of features Δ called a *feature set*. Then, for two objects a and b, A is the set of features in Δ possessed by a and B is the set of features in Δ possessed by b. Finally, one needs to choose some weighting function $f(\cdot)$, which is formulated over $P\Delta$. This allows one to write the following equality for the similarity between a and b (Weisberg 2012, 788):

$$S(a, b) = \Theta f(A \cap B) - \alpha f(A - B) - \beta f(B - A).$$

The equation says that the similarity between a and b is a function of the features shared by A and B subtracting the features not shared by A and B. Yet, unlike the Schelling model, actual cities are typically not perfectly grid-like and real people might not do their deciding based on an idealized utility function.

The same holds for Shagrir's (2012, 2022) account of structural representation. As Shagrir states: the "isomorphism requirement holds only as an *ideal*" (Shagrir 2012, 521; emphasis added). Crucially, then, this means that the shared structure between a model and a target system, or between what does the representing and the domain being represented, is "a matter of *approximation*" (Shagrir 2012, 521; emphasis added). This is also true of the Schelling model. Shea (2020) also takes the instantiation of structural correspondence to be an approximate instantiation. Shea (2020) considers a representation of relation H on a set of entities x_1 to x_4, where H represents the hierarchical relation of dominance between a group of macaques. The representational vehicles are listed as v_1 to v_4. According to Shea, for "any mapping *I* of the v_i onto the individual macaques x_i, there is a relation V on the v_i that corresponds to H: to see if V obtains between two vehicles, map them to the corresponding individuals under *I*, x_i and x_j and see if x_i is above x_j in the hierarchy (i.e., see if H obtains between x_i and x_j)" (2020, 112).

Shea's account of structural correspondence has a lot more detail than this. Here I am going to deliberately ignore these details, since they are irrelevant to the point that I wish to highlight now. According to Shea:

> The definition of structural correspondence we have been using is an exact one. *That is an idealization* [emphasis added]. . . . Consider a structural correspondence *I* that maps co-activation to distance with a certain metric, and consider a certain co-activation delay V that occurs between the activity of two place cells v_i and v_j. *I* map these to locations x_i and x_j and map V to a spatial separation of 1 cm. We can say *I* is *approximately instantiated* on occasions when the actual distance between x_i and x_j is approximately equal to the distance to which V maps under *I*, in this case 1 cm. (Shea 2020, 140)

The claim that the nervous system approximates structural correspondence between relations of internal states and relations of external states does not imply that it is literally true that the nervous system makes use of structural representations. This was also the case in the Schelling model.

As illustrated by appeal to the Schelling model, there is a rich and complex literature in the philosophy of science on structural similarity or isomorphism. Why does a model represent a target system? Because the model and the target system share a certain structure. In the case of the Schelling model, there is a formal similarity between the model and the target system because the target system shares an essential feature of the model: the Schelling dynamic. However, the Schelling dynamic is an idealization, in the same way that Shea (2020) admits that his definition of structural correspondence is an idealization. What one can say is that the Schelling dynamic approximates residential segregation in actual cities, even if the Schelling dynamic is an idealization. The same should be said with respect to structural representation.

Finally, there are well-known problems with isomorphism accounts in both the philosophy of science and cognitive science. Egan notes that "isomorphisms are cheap" since a "given set of internal states or structures is likely to be structurally similar to any number of external conditions" (2018,

250; both quotes). In the discussion of scientific representation in the philosophy of science, Quine (1969) has argued that any general notion of similarity is fundamentally problematic, since one cannot explain it in terms of more empirically or logically basic notions. Quine concluded, on this basis, that the concept of similarity is "logically repugnant" (59), and that mature sciences should dispense with notions of similarity relations altogether. Chakravartty captures the idea that appeals to similarity are too weak a relation to assess the model-world relation through the following case:

> The positions and velocities of the earth and moon in the earth-moon system are very close to those of a two-particle Newtonian model with an inverse square central force. The earth and moon form, to a high degree of approximation, a two-particle Newtonian gravitational system. (Chakravartty 2001, 334; from Giere 1988, 81)

However, if all "similarity" means is that a model based on Newtonian theories of gravity enables the computation of values for positions and velocities that match our observations, then similarity is too weak a notion to substantiate realism; including realism about structural representation.

5.3 Idealization and Neural Representation

From chapter 2 onward, I have emphasized how model building is accompanied by two different kinds of modeling practices: abstraction and idealization.

Abstraction need not present any problems for passing the translation task. Recall, this is the task of showing that facts about scientific models can be turned into true statements about the target system. What Cao (2022) dubs the "probing method" is an example of one kind of method used to build simpler models in computational neuroscience. It is basically the idea that when one probes or intervenes in a system, it becomes possible to identify causally relevant features to explain some target phenomenon. Functional magnetic resonance imaging (fMRI) is a probe of this kind. It is used to infer patterns of activity that are functionally involved in certain target phenomena of interest. Abstraction does not present a problem for realism in the domain of science, including computational neuroscience. Where patterns of neural activity are inferred based on fMRI, one can make the claim that the probe picks out real patterns in the nervous system, as Cao (2022) says (pace Dennett 1991; see also Potochnik 2017). Probes enable neuroscientists to build simplified models of their target system.

Idealization is a modeling technique that can be argued to present a serious threat to any kind of realism in science, including computational cognitive science (chapter 4). There are good reasons to think that Cao's (2022) account of neural representation yields an idealization view of representation—in addition to the ones I alluded to in section 5.2.1. Here are two desiderata that Cao thinks should be met for a helpful notion of representation that do real explanatory work:

A. They should play a distinctive explanatory role. That is, they should be more than just causal intermediaries, but rather, play a representational role—in the sense of being content-involving, and not just computational in some syntactic sense.

B. They should bridge the mechanistic facts and the interpretation facts, somehow. That is, they should help us to understand the relationship between mechanistic facts and cognitive or mental interpretations. (Cao 2022, 21–22)

(B) can be handled by deflationary accounts of representational content (see, e.g., Egan 2018). Cao (2022) acknowledges this (see 2022, 18nn41, 44). The question is whether the account of neural representation has what it takes to back (A). If the assessment in this chapter is on track, the answer is no, since the probing method picks out patterns of neural activity that are causally related. Following Ramsey (2007), I argued that if it turns out that representations function as "mere causal relays," then the phenomenon of interest—the contentful, truth-conditional states—has disappeared. Hence, while representational content can be used to make sense of how to bridge between the mechanistic facts and the interpretation facts, there is no reason to think that facts about mechanistic description of patterns of neural activity reveal true claims about content-involving representations. This should be a reasonably expected outcome. The central idea of an interventionist account is that any claim about a causal relationship between two variables, *X* and *Y*, is a claim about how a particular value or probability distribution of *Y* would change as a result of an actual or potential intervention on the value of *X* (or the probability distribution of *X*) (Woodward 2003). This is one of the dominant views on explanation in the literature. However, the interventionist account of explanation itself is silent about

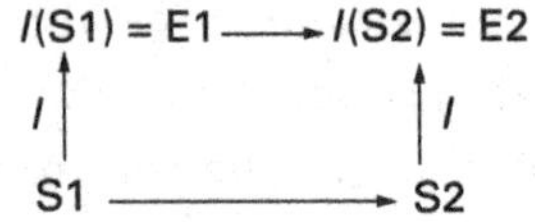

Figure 5.2
Modified version of Shagrir's (2022, 235) depiction of the information-processing description of the oculomotor memory system.

representation. It can be used to explain relations between the values of variables in diverse systems such as coupled pendulums, neural networks, and eco-systems.

One of Shagrir's (2022) cases concerns the oculomotor memory system. Through the lens of state-space analysis, it becomes possible to model this target system as shown in figure 5.2.

According to Shagrir (2012), "We can think of the memory system as consisting of states; each state, S_1, is a representation of a different eye position, E_1. The dynamics in the memory system can be seen as a transition from one state, say E_1, to another stable state, S_2, which represents another eye position, E_2, by computing some function . . . that operates on inputs that indicate change in eye position" (2012, 528; see also Shagrir 2022, 235). Figure 5.2 is an obvious idealized model, in the sense of representing state transitions as discrete states. Consider, for example, what Spivey says: the claim that a system "was in a particular 'state', **X**, at a particular point in time, really boils down to saying that the *average* of the system's states during that *period of time* was **X**" (2007, 30; emphases in original). Hence, to model representational states in terms of a discrete-state paradigm is to idealize the target system under investigation. The information-processing model depicted in figure 5.2 is an example of a model that provides a coarse averaging measurement of a system. This may be a practical necessity due to computational limits, but "should not be mistaken as genuine evidence for the system actually resting in a discrete stable state" (Spivey 2007, 30). Hence, the model of the oculomotor memory system is an idealization of its target domain.

5.4 Implications for Idealizing Neural Representation

I now consider three implications for realism about models of neural representation formulated against the backdrop of the RTM. My aim is not to defend any of these options. Rather, it is to present a series of illustrative options for thinking about neural representation in terms of idealizations—which I do based on the nonliteral view developed in chapter 3.

5.4.1 Representation: Epistemically Valuable and Ineliminable

The permanent view of idealization holds that at least some idealizations play a substantive role in science in the sense that they are indispensable for scientific progress. When given a nonliteral interpretation, the permanent view of idealization asserts that discourse about idealization should be taken nonliterally, is epistemically valuable, and ineliminable. Idealization, therefore, undermines the view that only the true, working parts are explanatorily significant. Note that the claim is not that one cannot arrive at true claims about the world by using this view of idealization. It is rather that because the model makes use of indispensable idealizations, the model itself will at best approximate some intended target system.

Here is the problem for realism about neural representations insofar as these posits are idealizations of this kind. One can employ a representational explanation for purposes of computational and/or cognitive tractability. In other words, one can use a representational explanation because scientists (and philosophers) are cognitively limited, and their subject matter is complex. Cao (2022) and Shea (2020) both argue that a distinct merit of representational explanation is that it makes a model of a cognitive capacity more tractable than a nonrepresentational alternative. Idealizations can do this work even if they are not literally true with respect to their intended target systems. As I argue in chapter 4, limit functions do not exist outside mathematical formalisms, for example. However, they have epistemic value. The same goes for point objects, infinite populations, frictionless planes, ideal rational agents, variational methods, and so on. What follows? Entity realism cannot be defended about idealized posits, unless one is willing to say that there literally are point objects, infinite populations, and so on. The same goes for the construct of neural representation. However, these posits can figure into a version of realism about scientific models as long as they are useful for scientific progress.

5.4.2 Representation: Epistemically Valuable but Eliminable

The nonpermanent view of idealization holds that at least some idealizations play a role in science but only in the short term. Idealizations are ultimately eliminable for theorizing about reality. When given a nonliteral interpretation, the nonpermanent view of idealization asserts that discourse about idealization should be treated nonliterally, is epistemically valuable but ultimately eliminable. In chapter 2, I referred to this view of idealization as Galilean idealization (after Weisberg 2007).

Consider how Galileo introduces an experimental idealization when discussing gravitational acceleration in the absence of a medium free of resistance. Galileo writes:

> We are trying to investigate what would happen to moveables very diverse in weight, in a medium quite devoid of resistance, so that the whole difference of speed existing between these moveables would have to be referred to as inequality of weight alone. . . . Since we lack such a space, let us (instead) observe what happens in the thinnest and least resistant media, comparing this with what happens in others less thin and more resistant. (Quoted in Weisberg 2007, 641)

Galilean idealization looks to be a promising strategy by which to defend a version of scientific realism. Or it looks more promising if the overall aim is to piece together realism, truth, and idealization—but only if, as I argued in chapter 4, Galilean idealization is treated nonliterally.

Idealization of this kind makes appeals to neural representation useful in the short run. The ultimate aim is to model the brain and mind in the absence of such theoretical constructs. An obvious pushback to this claim is that the nervous system is simply too complex to justify a version of Galilean idealization about neural representation. For instance, Chirimuuta writes:

> If the task is to explain how neural responses in sensory cortex enable a mouse to locate food, and how activity in motor cortex enables a mouse to move its body in the direction of that food, the key component of the explanation is the relationship between distal items—the smell of something edible, the layout of the territory, locomotion in the direction of food—and the cortex. The detailed causal mechanisms of olfactory reception, afferent and efferent signaling to and from the cortex, and neuromuscular activation can be relegated to the background to allow focus on the now more significant connections between brain, body, and environment. (Chirimuuta 2024, 160)

As Chirimuuta sees it, the function of representational discourse in neuroscience is to sharpen our focus on the significant connections between brain, body, and environment by downplaying the importance of providing "a detailed causal mechanism of olfactory reception, afferent and efferent signalling to and from the cortex, and neuromuscular activity" (2024, 160).

This is one of the central functions of idealization: to sharpen our focus on a particular set of causal factors, by setting aside many of the underlying causal complexities in the target system. Does this show that discourse about neural representation is permanent? Not at all. Once we treat appeals to mental or neural representations as idealizations, it is easy to motivate a preference for nonrepresentationalism: appeals to mental or neural representation can and should be eliminated because they are not explanatorily useful (i.e., they do not cut any interesting explanatory joints). Unlike cases such as the ideal gas law, where setting the value of intermolecular interactions to zero flags that this kind of behavior is explanatorily negligible, downplaying the importance of "a detailed causal mechanism of olfactory reception" does not highlight that neural representations are explanatorily important. If framed in terms of the FvI, neural representations are false theoretical posits. Thus, the FvI renders neural representational talk explanatorily problematic. Hence, appeals to neural representations do not actually sharpen our focus on the significant connections between brain, body and environment. Galilean idealization has the implication that entity realism about neural representation cannot be defended. However, one can defend a version of scientific realism about models that make use of such theoretical constructs (in the short run)—it requires only that one admits that constructs such as mental or neural representation are not actually true about the brain.

5.4.3 Representation: Epistemically Valuable but Causally Irrelevant

The final option for thinking about idealization is captured by Strevens's (2008) discussion of the ideal gas law. Briefly, it flags idealizations as falsehoods, causally irrelevant yet of epistemic value, since idealization assists with the identification of causal difference-makers in explaining the Boylean behavior of a gas. Here I make use of Strevens's account through my preferred nonliteral view of idealization. Strevens's (2017, 2019) view of idealization is embedded with an interventionist model of explanation. So is Cao's (2022) account of neural representation, given the focus on probing techniques in neuroscience. If representational posits are idealizations in the sense of Strevens (2017, 2019), the following obtains: flagging neural representations as idealizations identify representational posits as non-difference-making posits. This outcome goes directly against any view that takes the appeal to representations as making an explanatorily important difference.

On this view, then, representations are useful because they enable scientists to focus on the actual and relevant causal factors contributing to the production of the target system they want to explain. Consider some cognitive capacity you might want to explain—rat maze navigation, perhaps. An explanation will say something about the relevant neural patterns of activity, even if not everything that can in principle be said is captured in that explanation. The basic idea is that only those aspects that have a difference-making effect are relevant to the phenomenon to be explained. If neuroscientists are interested in explaining the wiring-and-connection facts that (arguably) realize cognitive capacities, then those facts are the relevant difference makers—the representational posits will not play any causal-difference-making role. They might have a pedagogical role to play, even if they make no causal difference. If this is the case, then one could still defend a version of scientific realism about models that make use of representational constructs. However, unlike the ideal gas law, where the idealization has real explanatory value, it is hard to see any reason for not eliminating discourse about neural representation entirely.

5.5 Conclusion

This chapter has treated the representational theory of mind (RTM) as a case of scientific model building. I showed that doing so invites a distinction between realism and antirealism about neural representation. Realists are committed to the claim that it is literally true of target systems that they manipulate physical vehicles with representational content. On this basis, I argued that, despite the realist commitments of the RTM, or some version of it, realists about neural representation face a problem. They fail to meet the translation task, the task of showing that it is literally true that the representational features of a cognitive model correspond to properties in the target system. What makes the translation task hard to address for many defenders of the RTM is that they admit that they are working with idealized (i.e., hypothetical) scenarios. No one claims that idealized posits have actual existence in nature. So one should not claim that idealized scenarios imagined in cognitive science have actual existence in nature.

6 Idealization and Content in Computational Modeling

Overview: In this chapter, my focus is on Egan's (2018) account of the nature and function of content in computational models. Egan's (2018) view is that semantic properties have some heuristic value only in enabling us to think about the behavior of cognitive systems in a useful way. Egan states: "The pragmatic elements and the contents they determine [read: the cognitive content] are 'quarantined' in the intentional gloss, to use Mark Sprevak's (2013) apt expression" (2018, 256). Egan labels semantic properties "cognitive content." I take Egan's treatment of cognitive content to suggest that content ascription in cognitive explanation is akin to idealization in science—as argued in the previous chapter. Egan argues that representational content should be seen as ineliminable parts of cognitive science; yet, Egan rejects the view that representational content captures anything real in cognitive systems. In this sense, Egan aligns her view on representational content with fictionalism—the idea that discourse about semantic content is useful but ultimately false. This is how idealization is defined under the standard view of idealization (see chapter 2).

A different way of describing Egan's position is by introducing Dennett's (1987) distinction between realism and instrumentalism. Unsurprisingly, Dennett is a realist about brains and their various neurophysiological parts. Yet, he is an instrumentalist about belief-desire states that, he contends, appear as abstracta in the "intentional stance." As Dennett states: "They are reason-giving explanations, which make an ineliminable allusion to the rationality of the agent. Primarily for this reason. . . . I suggest that folk psychology might be best viewed as a rationalistic calculus of interpretation and prediction—an *idealizing, abstract,* instrumentalistic interpretation method" (1987, 48; emphasis added). My suggestion is that

Egan (2018, 2020) shares this view of Dennett's; although with reference to semantic content ascription at the subpersonal level. In other words, Egan's use of idealization is about the alleged semantic (representational) properties ascribed by many to the nervous system. Egan's realism concerns the content of what she call the "computational theory of cognition proper"—the mathematical content of computational models in cognitive science.

In this chapter, I argue that the mathematical content that specifies a computational model is an idealization, and that it specifies only the computational model, not the target system. If treated in terms of the FvI (chapter 2), Egan's account is false. It is false because it ascribes idealized (false) properties to the target system (the brain). However, if given a nonliteral treatment (chapter 3), Egan's proposal about mathematical content in computational cognitive science is in a position to provide at least approximately true explanations of target systems. This, I think, is the real pay-off of the nonliteral view of idealization: that the theoretical posits of scientific theories that do some of the explanatory work do not come out as falsehoods. Although my focus is on Egan's account of the mathematical content in computational cognitive science, it applies to any case where a formal model of computation is inverted to explain the underlying dynamics of neural systems (see chapters 7 and 9).

6.1 Introduction

The construction of computational models is standard practice in computational neuroscience. Many take computational models to be representational models, given the assumption that cognitive activity is the acquisition, transformation, and use of representational content. This commits computational models to a distinction between a representational content, on the one hand, and a physical, representational vehicle, on the other hand. Following Ramsey (2023), I noted in chapter 5 that the distinction between content and vehicle is likely itself to be an idealization.

Despite the ubiquity of positing content to explain cognitive capacities, there is little consensus on how precisely the nature and function of content are best understood (see also chapter 5). Yet, there is at least some agreement that the following three conditions must be accounted for in order for a computational model to be explanatorily robust: (1) it must establish how content can be *determinate* (e.g., that the content of a frog's prey-detection system is "fly," and not something else); (2) it must allow for *misrepresentation* (i.e., it must be possible for a system to get things wrong); and (3) it must respect the *naturalistic constraints* on model building (i.e., it must minimally seek to specify its intentional or semantic content in non-intentional or nonsemantic terms).

Egan (2018) starts by looking at the most promising naturalized theories of content in the field: (1) information-theoretic approaches (e.g., Dretske 1986); (2) teleosemantic accounts (e.g., Millikan 1989; Shea 2020); and (3) structural-similarity views on content (Shagrir 2012). Egan argues that they all fall short of providing an adequate account of content; they all fail to meet, in different ways, the conditions for explanatory robustness. In chapter 5, I argued that one significant reason for this is that all these representational models are idealized models (see also Coelho Mollo 2021). Egan's proposal for generating a proper understanding of the nature and function of content in computational models is what she calls a *deflationary account of content*—an intermediary position between "industrial-strength" realism and content eliminativism.

In this chapter, I focus primarily on Egan's (2018) account of mathematical content in computational cognitive science. Egan's (2018, 2020) view of content in computational models can be seen as an attack on "the received view" of computation, according to which a necessary condition on a physical process qualifying as a computational process is that it has semantic content (Shagrir 2022; Sprevak 2010). Of course, if it is correct to identify discourse about internal states with semantic content as idealizations, then the received view is false from the start. Egan's proposal is that computation can, and should, be understood in mathematical terms only—that is, semantic content is not part of the instantiation of a computational process. The view defended in chapter 5 supports this conclusion, at least insofar as semantic content ascription is a way of idealizing a target system. Sprevak (2010) aims to show that Egan's (2018) attack on the received view fails. I have a different aim: to assess the relationship between a computational model formulated entirely in mathematical terms and its intended target system (the nervous system). Specifically, I argue that the mathematical content posited to explain computational models fails to correspond with the properties of real-world target systems. In other words, the mathematical theory of computation, as formulated by Egan, confuses the properties of the computational model with the properties of the target system the model is used to understand. A different way of putting this is: Egan succeeds in showing only that mathematical content ascription is literally true of the computational model, not the target system.

I proceed as follows. In section 6.2, I motivate Egan's (2018) distinction between two notions of content: cognitive content and mathematical content. In section 6.3, I introduce the mathematical theory of computation in the context of computational neuroscience. Note that unlike people like Piccinini (2015) and Shagrir (2022), Egan is not attempting a theory of computation. Egan's aim is to provide an account of the content of a computational model and its role in computational explanation. I argue that if the aim is to justify a version of the computational theory of cognition (CTC), then one cannot treat the computational model as literally true of the target system. The problem with Egan's account is that the mathematical content sits on the side of the scientific model. It does not sit on the side of the target system. There is no problem with this in principle. The problem resides solely in the realist claim about what defines content in the computational theory of cognition proper (as Egan calls it). I conclude in section 6.4.

6.2 Two Kinds of Content: Cognitive and Mathematical

A strategic decision has to be made when engaging in cognitive modeling. What level of analysis should one adopt to tackle questions about cognition? A traditional theory of multilevel analysis asserts that there are different levels on which one can investigate cognition, each of which involves different degrees and kinds of detail. An influential version of multilevel, or hierarchical, computational explanation is Marr's (1982) three-level model of analysis (see also chapter 8). At the top, there is the *computational level*, where the modeler needs to determine the computation to be performed to explain some cognitive activity. In the middle is the *algorithmic level*, where the modeler must figure out by which process the computation, specified at the top, is carried out. Finally, at the bottom, there is the *implementational level*, where the goal is to determine or identify the physical realizers of the algorithm determined in the middle of the hierarchy. Sprevak (2010) notes that Marr's (1982) computational level has been equated with Pylyshyn's (1984) semantic level, Newell's (1982) knowledge level, and Dennett's (1987) intentional level. Egan's (2018) mathematical theory of computation attempts to show that identifying the computational level of Marr's (1982) three-tier hierarchy with representational content is a mistake.

According to Egan, the following two features are important for computational theorizing:

1. Computational theories of cognitive capacities provide what I call a *function-theoretic* characterization of the capacity.
2. A computational theory (including the function-theoretic characterization, and the specification of algorithms, structures and processes) is accompanied by what I call an *intentional gloss*. (Egan 2018, 251)

Egan accompanies (1) and (2) with a distinction between two kinds of content: mathematical content and cognitive content. Here we are told that the functional-theoretic characterization of some capacity is captured by specifying the mathematical content of a computational model. Moreover, Egan says that the mathematical content used to specify a computational model can be accompanied by cognitive content, where cognitive content is quarantined in an intentional gloss on the target system. As Egan puts it:

> Cognitive contents . . . are not part of the essential characterization of the device and are not fruitfully regarded as part of the computational theory proper. They are ascribed to facilitate the explanation of the relevant cognitive capacity. (Egan 2018, 253)

For Egan, cognitive or semantic content plays only a kind of instrumental epistemic role, in the sense that semantic content facilitates scientific understanding of how a target system can perform a certain kind of cognitive activity. Unlike standard naturalistic theories of representational content, where content is said to be determined by a privileged relation holding between an internal state and the external state it is about, content ascription by computational theorists, Egan argues, is specified only in ways that are salient for explanatory purposes. Why does my dog chase the neighbor's cat? Here pragmatic considerations might likely privilege the distal cause (the cat in my backyard) rather than a proximal cause (e.g., certain patterns in the retinal image). Yet, for Egan, this relation is not determinate. Content is ascribed if it enables or facilitates an explanation of why my dog is rushing across the yard in pursuit of a cat. If the goal, however, is to explain my dog's visual mechanism, then the content ascription "dark patterned moving object" is likely to be preferable to "cat." Hence, on this view, different contents can be epistemically valuable in explaining how a target system is able to perform one and the same task depending on the kinds of questions scientists are interested in. For Egan, then, representational content is just a gloss on the system. It is useful for the theorist to make sense of the behavior of a target system. However, representational content—on the side of the model—does not have any mind-independent status on the side of the target system. This is also what Egan concludes here: "This paper sketches an account of the nature and function of representation in cognitive neuroscience that couples a *realist construal* of representational vehicles with a *pragmatic account* of representational content" (2020, 26).

The pragmatic account of representational content is consistent with a version of scientific realism as I developed it in chapter 4. The only price to pay is that one cannot treat talk about inner states with representational content literally. If mental or neural representations are idealizations, then it is, after all, not the case that nervous systems instantiate semantic content. So, to defend scientific realism in this context requires that one acknowledges that semantic content ascription is a way of idealizing a system—for different explanatory purposes depending on the questions of interest about some target system. If taken literally, representational descriptions of target systems are false in the sense that representational descriptions do not cut any objective, mind-independent joints. This is very close to Egan's (2020) view. Consider what Egan says here: "The choice of content will be responsive to such considerations as ease of explanation, and so *may involve considerable idealization*" (2020, 34; emphasis added). On Egan's (2018, 2020) view, one can accept the use of representational content, and one can accept that this has epistemic value even if representational content is quarantined in the intentional gloss.

The realist account of the representational vehicles is Egan's (2018) account of mathematical content in computational cognitive science. So, we can ask: In what sense can one defend realism about the CTC on the basis of having to translate a description of a computational model into truth-conditioned assertions about a target system? This is the translation task introduced in chapter 5. The translation task—from a model to a target system—requires that one can arrive at a realist picture about X only if one is able to specify how X and its properties, $X_1 \ldots, X_n$, can be turned into truth-conditional ascriptions of the relevant target system in question. You might think this demand is too high. For example, one option that looks to justifiably circumvent the translation task would be to ascribe representational features of internal states of cognitive systems based on how well they explain the sorts of patterns of behavior one might be interested in explaining (see, e.g., Pylyshyn 1984). By adopting this option, one could argue that representational posits capture, even if with some imprecision and approximation, features that cognitive systems actually—or literally—possess. One could try to circumvent the translation task in this way. Yet, this option comes with a significant cost. It is entirely consistent with a version of neural fictionalism about representational posits in cognitive explanations. To see this, note that we do not have direct access to the neural states underlying overt behavior. One can probe the brain with various techniques and tease out patterns of neural activity. Unless one can specify which aspects of neural activity correspond to which aspects of a specific target system, then the fictionalist will always have the option of arguing: we do not know if *X* is true or false. However, given that *X* is useful for explaining the pattern of behavior we are interested in explaining, we can treat *X* as true in the fictional story about *X* (Sprevak 2013). Note that it will not work to say: given that representational features are substantially constrained by both explanatory and predictive considerations, those considerations seem hard to motivate in the absence of realism. It will not do to say this because this specific option is equally open to the fictionalist about neural representation. Fictionalists can agree that neural representation talk enables us to explain cognitive capacities, even if such talk is not literally true of the target system under investigation.

Egan's (2018) proposal faces the same challenge, but about the mathematical content of a computational model. That is, here the challenge is to show how the mathematical content used to specify a computational model can be translated into truth-conditional assertions about the target system.

6.3 The Mathematical Theory of Computation for Cognition

Mathematical modeling is a pervasive research strategy in computational cognitive science. The HH model is an example (see chapter 1). A different example is the first mathematical model of a neural network developed by McCulloch and Pitts (1943)—also described in chapter 1. Egan (2018) calls the aim of explaining cognition via mathematical modeling *function-theoretic explanation*, and she calls the mathematical characterization that is central to this strategy the *function-theoretic characterization*. Egan draws explicitly on Marr's (1982) theory of vision to present her account of mathematical content in computational models. Like Marr (1982), Egan notes that theories of early vision seek to explain processes such as edge detection, and they do this partly by modeling this process as involving a specific computation: taking the Laplacian of a Gaussian distribution of the retinal array. A different case Egan (2020) considers is recent developments in Bayesian modeling. In particular, predictive coding models seeking to explain cognition in terms of prediction error minimization or belief optimization.

Central to predictive coding models is the idea that the internal states and active states of organisms can be modeled as computing Bayes's rule to optimize the posterior distribution of their beliefs. Bayes's rule states:

$$P(A \mid B) = \frac{P(B \mid A) \cdot P(A)}{P(B)}.$$

$P(A|B)$ represents the posterior probability of A given B. Bayes's rule says that one can calculate the posterior probability by multiplying the conditional probability of B given A multiplied by the prior probability of A divided by the probability of B (also known as model evidence). For example, say we want to figure out the posterior probability of someone having a disease A conditioned on symptoms B. One way of doing so would be to use Bayes's rule: take the conditional probability of B given A; then multiply the conditional probability with the prior probability of the disease being A divided by the symptoms B.

For Egan (2018), a function-theoretic chacterization is an abstract, domain-general, and content-free characterization of the function-theoretic explanation of target systems in the world. As Egan states: "It prescinds not only from the cognitive capacity that is the explanatory target of the theory (vision, motor control, etc.) but also from the environment in which the capacity is normally exercised" (2018, 252). In this way, even the function-theoretic characterization is an idealization.

What is the relation between a function-theoretic characterization of a computational model and its target system? I want to suggest that if we take this kind of characterization of a system literally, then the claims about function-theoretic explanation are false. Or, the function-theoretic characterization of organisms optimizing their beliefs via the computation of Bayes's rule is false, if understood literally. Let us consider S again, where S is = "p and p_1 . . . and p_n and J" (from chapter 4). S is the claim that nervous systems engage in processes that minimize uncertainty and that nervous systems optimize their beliefs in light of new evidence and that nervous systems compute Bayes's rule. *J* is the claim that nervous systems compute Bayes's rule. If *J* is given a literal treatment, then *S* is false. However, if *J* is treated nonliterally, that is, as neither true nor false, then it is open for us to explore the extent to which *S* can be said to be at least partially true—or approximately true—with respect to its target system.

The same holds for taking edge detection in early vision as the result of the visual system computing the Laplacian of a Gaussian distribution. Laplacian filters are also known as derivative filters, and are often used to find areas of rapid change (or edges) in images. Derivative filters have a high sensitivity to noise. In practice it is therefore common to use a Gaussian filter to smoothen the image (or reduce the noise). This two-step process is called computing the Laplacian of a Gaussian. In Egan's words:

> Marr's (1982) theory of early vision purports to explain edge detection, in part, by positing the computation of the Laplacean of a Gaussian of the retinal array. The mechanism takes as input intensity values at points in the image and calculates the rate of intensity change over the image. In other words, it computes a particular smoothing function. (Egan 2018, 252)

The Gaussian filter is mathematically expressed as (Zhang et al. 2020, 184):

$$G_\sigma\ (x,y)=\frac{1}{\sqrt{2\pi\sigma^2}}exp\left[-\frac{x^2+y^2}{2\sigma^2}\right],$$

where x is the distance from the origin in the horizontal axis, y is the distance from the origin in the vertical axis, and σ is the standard deviation of the Gaussian distribution. According to Zhang et al., an original image $f(x*y)$ is processed by a convolution operation with a Gaussian kernel $G_\sigma\ (x,y)$ and the Laplacian calculation is performed to suppress the noise of the image (Zhang et al. 2020, 184):

$$\Delta[G_\sigma\ (x,y)*f(x,y)]=[\Delta G_\sigma\ (x,y)]*f(x,y)=LoG*f(x,y).$$

Egan (2018) explicitly states that the account of representational vehicles, formulated entirely based on a mathematical theory of computation, is a realist account. However, the function-theoretic characterization of edge detection in early vision in terms of computing the Laplacian of a Gaussian is an idealization. This is easy to see. In an ideal world, visual input is not noisy. Visual input would greet the visual system as smooth, regular, clearly marked, and with sufficient density of detail "so that gradients in an image can be measured quite precisely" (Marr 1982, 233). Applying the Laplacian to a Gaussian distribution allows one to create this kind of ideal world mathematically. If the function-theoretic characterization of edge detection in early vision in terms of computing the Laplacian of a Gaussian is treated literally, the account is false. By implication, therefore, the function-theoretic characterization of edge detection in early vision in terms of computing the Laplacian of a Gaussian must be understood nonliterally. In other words, one can defend a version of scientific realism about Egan's mathematical theory of computation but only if one treats the idealization—that edge detection in early vision is computed by taking the Laplacian of a Gaussian distribution—nonliterally.

The moral of the story is: if one wishes to justify scientific realism about the use of computational modeling to understand cognition, one cannot treat the computational model as literally true of the target system. This would be to confuse the map with the territory. As Andrews (2021) puts it in a different context: "The math is not the territory" (2021, 1). For the same reasons that we should not infer from the Hardy-Weinberg model that biological populations are infinite, we should not infer from computational models of early vision that edge detection is literally computed by taking the Laplacian of a Gaussian distribution. Or, one should not infer on the basis of a Bayesian model of the nervous system that the brain is literally a Bayesian machine.

6.4 Conclusion

It is tempting to see computational descriptions of scientific models as mirror images of how target systems succeed in doing what they do. After all, the model is invoked to explain aspects of target systems. In this chapter, my specific focus has been on Egan's (2018) view of the mathematical content that describes a computational model and which Egan deems to constitute the computational theory of cognition proper. I argued that the mathematical content that specifies a computational model specifies only the computational model. It does not specify the target system. Or it does not specify the target system literally. Alternatively, if we proceed on the assumption that it does map on to the target system, the claim is false. To block this conclusion, I argued that Egan's proposal can be argued to be approximately true of target systems but only if the mathematical content specifying the computational model is treated nonliterally.

7 Idealization and the Mechanistic Theory of Computation

Overview: This chapter focuses on Piccinini's (2015, 2020) influential mechanistic theory of computation. A key idea behind the mechanistic theory of computation is that a computational explanation of a system takes the form of mechanistic explanation. Another central idea that motivates the mechanistic view of computation is that computation involves three elements: (1) rule, (2) medium-independence, and (3) teleological function. Hence, the mechanistic view of computation can be applied in a defense of the CTC, given the assumption that cognitive processes involve manipulation of medium-independent vehicles, whose (teleological) function it is to carry representational content in accordance with rules mapping across inputs, internal states, and outputs.

The central point that I defend in this chapter is: the claim that the nervous system and cognition are computational is false if taken literally. Maley (2023) argues that computational processes are not executed over medium-independent vehicles. I will not address this issue here. My target is the specific condition that computational processes in the nervous system are done in accordance with a rule. I shall show that the "rule-governing" condition on computation is not reflected in the target system. Rather, it sits entirely on the side of the scientific model (or, the computational model); not the nervous system. Again, we will have reasons to think that this can be made consistent with a version of scientific realism. However, once the rule-governing condition cannot be met, realism about the brain and cognition being computational cannot be defended.

7.1 Introduction

An advantage of the mechanistic account of computation is that it does not rely on question-begging notions from logic and computer science such as algorithm, program, automate, formal proof, and so on (Shagrir 2022, 148). Instead, it appeals to the mechanistic view of explanation in the philosophy of science. Indeed, if any perspective in philosophy of science and neuroscience can lay claim to being dominant, it is the mechanistic one (Chirimuuta 2014, 128).

There are two parts to Piccinini's (2015, 2020) mechanistic theory of computation. The first is physical computation and the second is the relationship between physical computation and information. I say that there are two parts because physical computation does not, according to Piccinini (2020, 129), imply any kind of information processing. Or, more specifically, physical computation does not entail representation. This sets Piccinini's (2020) account of physical computation in sharp contrast to someone like Fodor, who famously argued that there can be "no computation without representation" (1975, 34). Here is Piccinini's official statement of his position: "Physical computation is the processing of medium-independent vehicles by a functional mechanism in accordance with a rule" (2020, 155).

Here is how I shall proceed. In section 7.2, I focus on the link between computation and mechanistic explanation. I shall argue that the claim that the brain is computational qua mechanistic is an idealization, since modeling in computational neuroscience is often carried out by appeal to computational models that are nonmechanistic. This situation is similar to what I discuss in chapter 8; namely, that science often proceeds to study one and the same target system by using many different simplified and idealized models. Piccinini (2020) acknowledges that simplification and idealization are often necessary when building mechanistic models of computational processes. In section 7.3, I discuss what makes a mechanistic explanation a form of computational explanation. Here I focus exclusively on the rule-governing aspect of the mechanistic theory of computation. I argue that the claim that the nervous system maps inputs, internal states, and outputs in accordance with a rule is an idealization. The result is that work in computational neuroscience does not provide support for the CTC. I conclude in section 7.4.

7.2 Computation and Mechanistic Explanation

Piccinini states the following about physical computation:

> The present account is mechanistic because it deems computing systems a kind of functional mechanism—mechanism with teleological functions. Computational

> explanation—the explanation of a mechanism's capacities in terms of the computations it performs—is a species of mechanistic explanation. (Piccinini 2015, 118)

For the purposes of exposition, I shall treat mechanistic explanation as a model-based view of science, where the aim is to build mechanistic models to account for certain target phenomena. Mechanistic modeling of a target phenomenon proceeds by decomposing a mechanism into its constituent parts in order to localize their characteristic activities, including the spatial and temporal relations between parts of a mechanism (Bechtel and Richardson 2010, 7).

The first modeling assumption of mechanistic explanation is that target systems are in fact decomposable. The second modeling assumption is that only causally relevant activities are included as constituent parts of the description of the mechanism, leaving aside any irrelevant factors. This is the same strategy adopted by Strevens (2008) in his discussion of the role of idealization in science. In short, for a description of a mechanism to explain its target phenomenon, it is required only to include the causally relevant parts of the mechanism responsible for the occurrence of the target phenomenon. As Piccinini says:

> According to the mechanistic account, computational explanation is mechanistic. That is, we explain computations by decomposing them into subcomputations, each of which is performed by computing components which, when organized in the right way, give rise to the whole computation. (Piccinini 2020, 149)

Let us assess the situation in light of *S*, where *S* is = "*p* and p_1 . . . and p_n and *J*." Let *S* state that "the nervous system is computational and that nervous systems have parts and that any computational explanation of the nervous system is mechanistic." I shall treat the last part of *S* "that any computational explanation of the nervous system is mechanistic" as *J*. Prima facie justification for doing so comes from the mechanists themselves. Here is what Bechtel and Richardson say:

> Pursuing decomposition and localization *is to impose an assumption* on the nature of the system whose activities one is trying to explain: it is assuming that it is decomposable. A decomposable system is modular in character, with each component operating primarily according to its own intrinsically determined principles. (Bechtel and Richardson 2010, 24–25; emphasis added)

If the modeling assumptions of mechanistic explanation—decomposition and localization—are imposed on the nature of systems, it raises the following kind of question: Are all natural systems decomposable, such that one can localize the activities of each part of a mechanism, where each part performs its own modular functions? This is far from obvious if the scope of the claim is that all natural systems are mechanistic systems. Shagrir suggests that a network model that describes patterns of traffic in a road network explains the amount of traffic in each road based on a set of intricate relationships across the whole network. Shagrir notes that the main reason that a certain road has lighter or heavier traffic "depends on the structure and organization of the entire network, and cannot be explained by decomposing the network into separate components and their organization" (2022, 154–155). This conclusion is similar to Huneman (2010), who observes that for some target phenomena, the best available explanation is not one that proceeds on the assumption that decomposition and localization can be imposed on the nature of systems in order to explain their behavior since such cases call out for explanation in terms of topological or network properties of a target system.

Let us restrict the scope of the claim to the nervous system. What follows? Piccinini (2020) holds the view that nervous systems are computational systems and computational systems are mechanistic systems. Hence, physical computation in the brain is mechanistic, proceeding via both decomposition and localization to an explanation of the relevant mechanisms constituting their respective target phenomena. According to the mechanistic view of computation, all computational processes in the nervous system are mechanistic. Does this follow? I said above that I treat *J* as the claim "that any computational explanation of the nervous system is mechanistic." I do so because several authors have pointed out that computational neuroscience, from where the CTC partly draws evidence for its justification, involves computational explanations that are not mechanistic (see, e.g., Shagrir 2012; Shapiro 2017; Weiskopf 2011).

Chirimuuta (2014) provides a detailed account of nonmechanistic computational explanation in computational neuroscience based on efficient coding models. Nervous systems must operate in an energy-efficient manner. Efficient coding models are based on the idea that one way nervous systems can do this is by adapting their internal and sensory states to the statistics of sensory inputs conditioned on their behavior. That is, the statistics of sensory stimuli can be directly shaped by an organism's actions in the world (see, e.g., Kirchhoff et al. 2018; Parr, Pezzullo, and Friston 2022). Chirimuuta (2014) follows Dayan and Abbott (2001) in observing that models in computational neuroscience are usually interpretive rather than phenomenal or mechanistic. She states:

> Interpretative models are central to explanations of why a particular neuronal type or brain area is organised in a certain way, explanations which typically make reference to efficient coding principles. In contrast, phenomenal models describe what is there, and mechanistic models explain how systems work. Information-theoretic and computational principles are central to interpretative modelling, and causal-mechanical descriptive detail and accuracy is not required. For this reason, most interpretative models are minimal. That is to say, they typically abstract away from many biophysical details of the neural system, in order to highlight dominant causal influences or universal behaviour. (Chirimuuta 2014, 128)

As the quotation reveals, even interpretive models are idealizations in the specific sense of sharpening attention on specific (and minimal) principles of neuronal dynamics.

Chirimuuta (2014) illustrates efficient coding explanations through a detailed discussion of Heeger's (1992) contrast normalization model of the primary visual cortex, which, as she notes, is viewed as a canonical neural computation. A general feature of any canonical neural computation is that models that exemplify instances of it are "described at a high level of abstraction from biophysical implementation" (Chirimuuta 2014, 129). In chapter 9, I discuss a different example of an efficient coding model that can be treated as a canonical neural computation: the FEP. The FEP is a mathematical model based on formalisms from statistical physics and machine learning. It seeks to explain why a certain neuronal type or brain area is organized the way it is by appealing to a universal function that any organism must reduce in order to survive: the minimization of entropy. The FEP does not contradict mechanistic explanation in the sense that one can proceed from the FEP to the study of neurons and their activities, as well as the relationship between cells and ensembles of cells. Furthermore, one can proceed from the FEP with the aim of discovering the implementation details for how entropy is minimized in a system. One can also use the FEP as a theoretical framework motivating the search for certain hierarchical configurations for neuronal activity that form the basis for entropy minimization in the brain. Nonetheless, it is not the mechanistic basis of such discoveries that explain why neurons or ensembles of neurons in the brain organize to minimize their free energy and, therefore, entropy. The explanation rests with universal principles for survival based on efficient coding principles. In other words, it is the imperative for minimizing entropy that explains neuronal dynamics, not the mechanistic details. The FEP is a mathematical computational model, which rests on an abstraction from biophysical details to aid understanding of the behavior of systems (Friston and Stephan 2007; Mann, Pain, and Kirchhoff 2022; Robertson, Kiverstein, and Kirchhoff 2024).

One way to present the issue is to say that there is a tension between the mechanistic framework and computational explanation (Shagrir 2022). Shagrir (2022) notes that "in his influential book *Explaining the Brain,* Carl Craver (2007) does not even mention the central role of computational approaches in the study of brain and cognitive sciences" (2020, 152). Chirimuuta (2014) observes that the contrast between computational and mechanistic explanations is commonplace in neuroscience.

We can conclude this section by making the following observations. The claim that any computational explanation of the nervous system is mechanistic is an idealization insofar as this claim is treated as literally true for all computational explanations of the nervous system. The idealization is an explanatory idealization, which claims that all possible computational explanations of nervous activity conform to the modeling assumptions of decomposition and localization. If this is the claim, then it is false, if taken literally.

7.3 What Makes a Mechanism Computational?

Piccinini's (2015, 2020) definition of physical computation involves three key elements: (1) rules, (2) medium-independence, and (3) teleological functions. Hence, physical computation is a type of teleofunctional mechanism, whose vehicles are medium-independent and where a rule maps from inputs (and possibly internal states) to outputs. According to Piccinini: "Computing systems are mechanisms whose teleological function is manipulating medium-independent vehicles in accordance with a rule. CTC should be understood to assert that neurocognitive systems are computing mechanisms in this sense" (2020, 207).

The structure of Piccinini's argument goes as follows (2020, 213):

1. Neurocognitive processes are functional manipulations of medium-independent vehicles in accordance with a rule.
2. Functional manipulation of medium-independent vehicles in accordance with a rule are computations.
3. Therefore, neurocognitive processes are computations.

Piccinini (2020) argues that the kind of vehicles manipulated by neuronal processes include things such as voltage changes in dendrites, action potentials (or spikes), neurotransmitters, and hormones. Piccinini observes that spike trains (see chapter 1) are the primary way neurons transmit long-distance signals to one another. Obviously, the kinds of vehicles manipulated by neuronal processes such as spike trains and their generation will depend, in some way, on the biophysical properties of the medium—the medium of spike trains, for example. However, as Piccinini (2020) notes, "The aspects of spikes and spike trains that appear to be functionally relevant to cognition are primarily spike rates and spike timing" (2020, 214). Crucially, spike rates and timing are properties of spikes and spike trains that can be realized across a number of different media. Hence, spike rates and timing are said to be medium-independent (see Maley 2023 for objections). Finally, as Piccinini argues, nervous systems manipulate vehicles such as spike trains in accordance with a rule:

> One particularly important rule is normalization, whose output is a ratio between the response of an individual neuron and the summed activity of a pool of neurons (Carandini and Heeger 2012). As Carandini and Heeger define it, normalization is a medium-independent operation found in many neural systems of many species, where it can be realized by many lower-level computations and can contribute to many broader computational and representational functions. Due to its ubiquity and versatility, Carandini and Heeger dub normalization a canonical neural computation. (Piccinini 2020, 214)

Let's consider *S* again, where *S* is = "*p* and p_1 . . . and p_n and *J*." *S* is the claim that "the nervous system manipulates neuronal vehicles and that certain properties of these vehicles are medium-independent and that the manipulation of vehicles is done in accordance with a rule." I treat the last part of *S*, "that the manipulation of vehicles is done in accordance with a rule," as *J*. One motivation for taking the rule-governing aspect of Piccinini's (2015, 2020) account of physical computation as an idealization is that Piccinini does not require that the system in question represent a rule to produce an output conditionalized on inputs and internal states. Piccinini states: "A rule in the present sense is a map from inputs (and possibly internal states) to outputs. It need not be represented within the computing system; it may be followed by the system without representing it anywhere within it" (2015, 121).

If the rule that governs state transitions within a system need not be represented by the system, where, then, is the rule? In part II, "Implications for Cognitive Science," I looked at the following examples: the amount of blood pumped by the heart per minute can be explained using the formula: Q [L/min] = HR [beats/min] × SV [mL/beat] (Hryvniak et al. 2021); and the acceleration of falling objects can be captured by Newton's law of gravitation: $F = G(m_1, m_2)/R^2$. I suggested we would not hesitate to conclude that the mathematical formalisms invoked to describe the behavior of both cases sit on the side of the model, not on the side of the target system. I made the same claim when assessing Egan's (2018) proposal that edge detection in early vision is computed by taking the Laplacian of a Gaussian distribution.

Piccinini (2020) notes that a particularly important rule is normalization, or divisive normalization. According to the authors cited by Piccinini, "Normalization computes the ratio between the responses of an individual neuron and the summed activity of a pool of neurons" (Carandini and Heeger 2012, 51). Normalization is often referred to as a canonical neural computation, as mentioned in the previous section. Other examples include exponentiation and linear filtering. According to Carandini and Heeger: "Linear filtering (that is, weighted summation by linear receptive fields) is a widespread computation in sensory systems. It is performed, *at least approximately*, at various stages in vision, hearing and somatosensation" (2012, 51; emphasis added).

Here is the central point: if a system approximates linear filtering, then it does not literally perform linear filtering. Similarly, if a system approximates Bayes's rule, then it does not literally compute Bayes's rule. And, if a system approximates the Laplacian of a Gaussian distribution, then a system does not literally compute the Laplacian of a Gaussian distribution. The system in question is modeled as transitioning from inputs (and possible internal states) to outputs in accordance with one of these computational rules. Yet, it does not follow that the system in question literally computes any of these rules. Indeed, if the rule need not be represented within the system itself, then the canonical definition of mechanistic computation is controversial for realism about the CTC. *J* is an idealization. It is false, if taken literally.

I will have more to say about approximation of formal computational methods in chapter 9. Here I want to note that Piccinini's (2020) appeal to rule-governed state transitions as part of his realist defense of the CTC suffers from the same problem as Egan's (2018) realist defense of her mathematical view of computational content: it confuses the computational modeling of a target system with the target system itself. One can defend scientific realism in the context of making use of computational modeling to understand aspects of the nervous system. However, if "rule-governing" is part of the definition of the CTC, then realism about the CTC cannot be sustained. As I said above, basically everyone working with a version of the CTC suffers from the same problem as Egan (2018) and Piccinini (2020): they confuse the model for the target system.

Chalmers (2011) makes the same mistake in formulating a causal account of computation. According to Chalmers (2011), a physical system implements a computation when the causal structure of a physical system mirrors the formal structure of a computation. Chalmers's preferred framework for defining computation more formally is combinatorial state-automata. Chalmers states:

> A physical system P implements a CSA M if there is a vectorization of internal states of P into components $[s^1, s^2, \ldots]$, and a mapping f from the substates s^j into corresponding substates S^j of M, along with similar vectorizations and mappings for inputs and outputs, such that for every *state-transition rule* $([I^1, \ldots, I^k], [S^1, S^2, \ldots]) \rightarrow ([S'^1, S'^2, \ldots], [O^l, \ldots, O^l])$ of M: if P is in internal state $[s^1, s^2, \ldots]$ and receiving input $[i^1, \ldots, i^n]$ which maps to formal state and input $[S'^1, S'^2, \ldots]$ and $[I^1, \ldots, I^k]$ respectively, this reliably causes it to enter an internal state and produce an output that map to $[S'^1, S'^2, \ldots]$ and $[O^l, \ldots, O^l]$ respectively. (Chalmers 2011, 329; emphasis added)

Formally, there is nothing wrong with this definition of computation in combinatorial state-automata. However, and in the case with Piccinini's (2020) appeal to rules, the rule-governed state transitions in Chalmers's (2011) account sit on the side of the computational model, not the target system. It is an idealization to describe the nervous system as computing a mapping from inputs, possible internal states, to outputs in accordance with a rule. Chalmers seemingly admits this when saying: "In the explanation of behavior, too, causal organization takes center stage. A system's behavior is determined by its underlying causal organization, and we have seen that the computational framework *provides an ideal language* in which this organization can be specified" (Chalmers 2011, 342; emphasis added).

The computational framework is, therefore, a useful idealization in the sense that it enables us to build sophisticated models of the nervous system in the form of artificial neural networks and cellular automata. However, it is false of nervous systems if taken literally.

To bring this point home, let us finish this section by considering the normalization rule, as described by Carandini and Heeger (2012), in the invertebrate olfactory system. The focus is on the fruit fly, *Drosophila melanogaster*. According to Carandini and Heeger (2012), the fruit fly senses odors via its receptor neurons, each of which expresses a single odorant receptor. The receptor neurons, in turn, project via spike trains to the antennal lobe—a region in the brain that is understood to be similar to the olfactory bulb in vertebrates. Carandini and Heeger (2012) note that the response R of an antennal lobe neuron increases with the activity I of the receptor that drives it. The following equation express the mathematical relation between these values (Carandini and Heeger 2012, 52):

$$R = \gamma \frac{I^n}{\sigma^n + I^n},$$

where the parameters, γ, σ, and n are said to determine the shape of the response curve. As Carandini and Heeger note: "Modeling γ and σ as constants is to treat these parameters in an *idealized fashion*" (2012, 52; emphasis added). The interaction of the parameters is provided by the normalization equation:

$$R = \gamma \frac{I^n}{\sigma^n + I_m^n + I^n},$$

where I is the response of the receptor neurons that drive the antennal lobe, and I_m is the pooled responses of the other receptor neurons. As Carandini and Heeger explains: "As I_m appears in the additive term in the denominator, increasing it has the same effect as increasing σ in [the equation above]: it shifts the response curve to the right on a logarithmic scale" (2012, 52).

My point is simple. The normalization rule is a computational rule that is invoked on the side of the computational model to *approximate* the dynamics on the side of the target system. This is perfectly sensible. Computational neuroscience could not proceed as it does without engaging with these kinds of modeling techniques. However, if one takes scientific discourse literally, then the claim that fruit flies compute certain outputs based on input and internal states by encoding and making use of the normalization rule is false. Hence, the rule-governing element of Piccinini's (2020) account of the CTC is on the model side of the model-world relation. Consequently, realism about the CTC can, therefore, not be defended. However, none of this implies that computational modeling cannot lead to real insights about the brain and cognition.

7.4 Conclusion

While one might be able to show that some forms of computation are mechanistic, there is evidence for the claim that not all kinds of computation are mechanistic—especially in the context of computational neuroscience. To insist otherwise is to impose a set of theoretical assumptions on target systems that they do not need to exhibit. Moreover, the claim that computations over input, internal states to outputs are done in accordance with a rule confuses the computational model for its target system. Models in computational neuroscience make use of idealized mathematical formulations to model the dynamics of target systems. However, one should not therefore conclude that the mathematical notations literally represent how the dynamics of target systems are carried out in the target systems themselves. See Chirimuuta (2024) for similar observations.

8 Idealization, Multiple Models, and Pluralism

Overview: In this chapter, I turn from specific proposals about the nature and function of cognition to *modeling practices* in cognitive science. The issue to be addressed is the prospects for *explanatory unification*: the idea that there is a single, unifying theory of mind that gets things right about the properties and dynamics of cognition. The central question is: should we expect, one day, to arrive at a single unifying theory of cognition? Or, should we expect that explanatory pluralism is the way forward in the sciences of the mind? As Milkowski and Hohol note: "The debate between the defenders of explanatory unification and explanatory pluralism has been ongoing from the beginning of cognitive science and is one of the central themes of its philosophy. Does cognitive science need a grand unifying theory? Should explanatory pluralism be embraced instead? Or maybe local integrative efforts are needed?" (2021, 1).

It is worth emphasizing that in many fields, it is common to make use of multiple different models of the same phenomenon (see, e.g., Weisberg 2013). Moreover, it is also standard practice to make use of the same model to understand different kinds of phenomena (see, e.g., Potochnik 2017, who discusses how game theory is being applied across different domains of inquiry such as evolutionary biology and behavioral sciences). Weisberg (2007, 2013) refers to the former use of multiple different models to explain the same target system as "multiple-models idealization" (MMI). MMI implies that the search for a single, unifying model of a target system is likely to be unsuccessful. If this is correct, then the aim of formulating a unifying theory of computation (say) is likely to be a misguided goal. Others take MMI to put pressure on veritism: that (approximate) truth is necessary for the epistemic acceptability of scientific theories and models.

Potochnik (2023), for example, emphasizes that the idea that science aims at truth is in tension with MMI. Finally, MMI can in many ways be seen as a motivation for explanatory pluralism, where pluralism between models can be thought of as integrative pluralism (Mitchell 2002), inferential pluralism (Hochstein 2023), or as a form of either perspectival realism (Chirimuuta 2018; Massimi 2022; Potochnik 2023) or deflationary realism (Coelho Mollo 2021).

Philosophers working on topics in computational neuroscience are increasingly attracted to explanatory pluralism, aiming to move away from explanatory unification. In this context, pluralism is being used to motivate different versions of scientific realism, such as deflationary realism and perspectival realism, both of which are argued to be justified by facilitating the construction of a better-rounded understanding of the brain and cognition.

In this chapter, I argue that the main worry about explanatory unification arises because MMI can be shown to imply the problem of inconsistent models (see, e.g., Morrison 2011). The problem of inconsistent models (the PIM) arises when multiple different and idealized models with incompatible features are applied to explain one and the same target system. In philosophy of science, the PIM is taken by some to pose a problem for scientific realism—realism in science is not compatible with the existence of mutually inconsistent representations of the same target system (see, e.g., Panagiotatou and Psillos 2023).

My aim will be to diffuse the PIM. I shall argue that the PIM is based on the fallacy of adopting a literalist stance on scientific discourse. The PIM only arises if one views multiple different and highly idealized models with incompatible properties as being literally true of one and the same target system. Crucially, if we can get rid of the PIM, then we can start with pluralism about scientific explanation and work our way to explanatory unification.

8.1 Introduction

Pluralism in science is no great surprise (Mitchell 2002). For example, since the ideal gas law was exported from thermodynamics to statistical mechanics, it has undergone various refinements, as expressed in the van der Waals equation. The van der Waals equation is a more complex gas model, since it includes both attractions and repulsions—the values of which are set to zero in the ideal gas law. Further refinements to the van der Waals equation led to the formulation of the virial equation, which can be rendered arbitrarily precise. The situation is that we have multiple models of gas behavior. As Mitchell notes:

> On scanning contemporary journals, books, and conference topics in some sciences, one is struck by the multiplicity of models, theoretical approaches, and explanations. Yet, if science is representing and explaining the structure of the one world, why is there such a diversity of representations and explanations in some domains? (Mitchell 2012, 55)

Mitchell (2002) draws a useful distinction between what she calls "competitive pluralism," on the one hand, and "compatible pluralism," on the other. Mitchell notes that one version of competitive pluralism is found in the work of Kitcher (1991). According to Kitcher, pluralism in the form of multiple competing theories is the most rational strategy to subscribe to given uncertainties surrounding which theories will turn out to be the most appropriate to adopt considering empirical evidence. The aim of this view, Mitchell says, is to resolve the conflict between the theories by settling on a single theory or model of some target system or domain. Mitchell quotes Kitcher, who states: "The community goal is to arrive at universal acceptance of the true theory" (1991, 19). Overall, the view presented here is that pluralism between competing theories is temporary and strategic, but "ultimately eliminable" (Mitchell 2002, 56). Compatible pluralism, as the name suggests, is where different theories or models work together by seeking answers to different kinds of questions and doing so by adopting different levels or scales of analysis. An example of this is Mayr's (1961) work in biology (discussed by Mitchell [2002]). Mayr sought to address two issues: (1) What kinds of questions can we ask? And (2) how should we answer these questions? In response to the first question, Mayr (1961) identified two different questions: how-questions, about proximate causation, and why-questions, about ultimate causation. Calcott provides the following illustration:

> For example, a request to explain sickle cell disease might elicit two very different responses. One response would include talk of how different proteins affect cell shape and how these in turn affect blood flow. But another response might, instead, refer to patterns of heredity in recessive genes. (Calcott 2013, 767–68)

Mayr's (1961) response to the second issue was to allocate each question to a different discipline. He argued that ultimate questions were questions best addressed by evolutionary biologists, whereas proximate questions were questions best addressed by those working in the field of functional biology. As Calcott notes: "Importantly, he presumed that these disciplines—and hence, the answers to each question—were largely independent" (2013, 768). Mayr writes:

> Any biologist working in one of these fields must have a knowledge and appreciation of the other field if he wants to avoid the label of a narrow minded specialist. Yet in his own research he will be occupied with problems of either one or the other field. (Mayr 1961, 1052)

This picture—which sees how-questions and why-questions as independent questions for different areas of science—has not gone unchallenged. Laland et al. (2013) reject the separation of disciplines proposed by Mayr (1961), arguing that answers to ultimate questions can and sometimes should incorporate details about proximate causes and mechanisms. Calcott (2013) argues that the dichotomy between proximate questions and ultimate questions is a false dichotomy, in the sense that one can ask diachronic questions about proximate mechanisms. I mention this here to frontload a discussion to come in this chapter: what seems to be two genuinely different ways of approaching one and the same phenomenon need not be two genuinely different and alternative ways of approaching the same phenomenon. Or, differently put, one might argue that computational and interpretive models in neuroscience are substantially different ways of understanding features of the same target system. My suggestion is that they are not. Similarly, as we shall see later in this chapter, one might argue that Bayesian models in neuroscience are substantially different from connectionist models. My suggestion is that they are not.

This goes against a general trend in the philosophy of science and philosophy of cognitive science, where it is being increasingly recognized that it is standard practice to make use of multiple different models to understand the same target system (Weisberg's [2007, 2013] MMI). MMI looks to be inconsistent with any form of explanatory unification. Or, this is how the implications of MMI are often interpreted.

Weisberg (2007) notes that unlike cases of Galilean idealization, where idealization is seen as temporary, MMI is permanent. Moreover, unlike Galilean idealization, where the aim is to arrive at a complete de-idealized and veridical model of a target system, MMI works counter to this aim. Perhaps explanatory unification is an unrealizable ideal? Weisberg notes that one frequently comes across instances of MMI in sciences dealing with extremely complex phenomena:

> In ecology, for example, one finds theorists constructing multiple models of phenomena such as predation, each of which contains different idealizing assumptions, approximations, and simplifications. Chemists continue to rely on both the molecular orbital and valence bond models of chemical bonding, which make different, incompatible assumptions. In a dramatic example of MMI, the United States National Weather Service (NWS) employs several different models of global circulation patterns to model the weather. Each of these models contains different idealizing assumptions about the basic physical processes involved in weather formation. Although attempts have been made to build a single model of global weather, the NWS has determined that the best way to make high-fidelity predictions is to employ all three models, despite the considerable expense of doing so. (Weisberg 2013, 103)

In Weisberg's analysis, MMI occurs when "our cognitive limitations, the complexity of the world, and constraints imposed by logic, mathematics, and the nature of representation, conspire against simultaneously achieving all of our scientific desiderata" (2013, 104). Potochnik (2017) argues that MMI is pervasive in science precisely for the reasons Weisberg mentions. Potochnik states:

> The effects of causal complexity and our cognitive limitations on science are quite general. Causal complexity makes it exceedingly difficult or impossible for a single representation to depict all of the significant causal influences, as is required for minimalist idealization. Our cognitive limitations, together with the world's complexity, make idealizations much more valuable than just as temporary expedients for our calculations, as is expected for Galilean idealization. (Potochnik 2017, 45)

The same sorts of observations are being made by philosophers of cognitive science. For example, consider what Sprevak and Colombo say in the introduction to their *Routledge Handbook of the Computational Mind*:

> Pluralist views about the computational mind are reflected in the progressive erosion of dichotomies that have traditionally defined the field. The contributions in this Handbook show that the historical roots of CTM are broad: even at its origins, CTM did not necessitate or reflect a monistic approach to the mind. Today, an increasing number of researchers realize that they do not need to pick between Turing machines, logic, neural networks, probability calculus, and differential equations as "the" approach to the mind. Nor do they need to claim that any one of these approaches is "the" computational theory of the mind. Instead, they

> are able to choose between diverse questions, problems, and approaches that fall within the computational sciences. These approaches have the power to illuminate different aspects of mental or neural phenomena. None has the monopoly on computation. This has led some researchers to reconceive of apparently competing approaches—such as connectionism or dynamical systems theory—as different aspects of the computational framework rather than as non-computational alternatives. (Sprevak and Colombo 2018, 3)

Conceiving of apparently competing models of the mind as lending different perspectives on various aspects of the brain (or cognition) is an example of MMI, where pluralism about the use of multiple models falls under the category of compatible pluralism. Given MMI, pluralism looks like it is here to stay. This sentiment is echoed by Weisberg (2007), noting that there is little to no search for a single, best model of predation in the ecological research program. He quotes Maynard-Smith, who states: "For the discovery of general ideas in ecology . . . different kinds of mathematical description, which may be called models, are called for" (1979, 1; quoted in Weisberg 2007, 656). As I mentioned in the overview to this chapter, some infer on the basis of MMI that the standard epistemic view of scientific realism is in tension with MMI (Potochnik 2023). Others argue that because of issues to do with compatible pluralism and problems surrounding veritism, MMI motivates versions of realism such as deflationary realism and perspectival realism (Chirimuuta 2018; Coelho Mollo 2022; Massimi 2022; Potochnik 2023).

In this chapter, I argue that MMI does not prevent explanatory unification. The main worry, or so I shall argue, about explanatory unification arises because MMI can be said to imply the PIM (Morrison 2011; see also Massimi 2021, 2022). The PIM arises when multiple different and highly idealized models with incompatible features are applied to explain one and the same target system.

Building on the work done across chapters 3–7, I argue that the PIM is based on the fallacy of adopting a literalist stance on scientific discourse. The PIM arises only if one views multiple different and idealized models with incompatible properties as being literally true of one and the same target system. Crucially, if we can get rid of the PIM, then we can start with pluralism about scientific explanation and work our way to explanatory unification. Furthermore, if we can get rid of the PIM, we can retain (approximate) truth as the aim of science. In other words, once the PIM is gone, MMI is no longer in tension with the epistemic thesis of scientific realism. Finally, once we get rid of the PIM, there is no need to adopt either deflationary or perspectival forms of realism.

I proceed as follows. In section 8.2, I motivate the use of MMI in more detail. In particular, I shall consider the relation between MMI and single-model integration, on the one hand, and MMI and multiple-model integration, on the other. In section 8.3, I review two examples of pluralism in science: Mitchell's (2002) account of integrative pluralism in biology and Hochstein's (2023) inferential account of integration in cognitive science. In section 8.4, I look at the relation between compatible pluralism and different versions of realism, especially perspectival realism and deflationary realism. In section 8.5, I show why deflationary realism can be seen as a form of inconsistent-friendly pluralism, which runs into the PIM. In section 8.6, I argue that perspectival realism, as it is used by Chirimuuta (2018) and Massimi (2022), can avoid the PIM. In section 8.7, I engage in a critical discussion of perspectival realism. I conclude in section 8.8.

8.2 Multiple-Model Idealization and Model Integration

Let us start with the semantic and accuracy questions concerning scientific representation introduced in chapter 1. The semantic question asks us to consider in virtue of what a model represents its target system (if at all)? The accuracy question is the question of what makes a model accurately represent its target system (if at all)? Consider the situations in which MMI becomes a relevant modeling technique—namely, in situations where more than one model plays a representational role for a target system. Massimi (2019) mentions a classic example from nuclear physics, where families of rival models for the atomic nucleus are routinely applied, such as the liquid drop model, the shell model, the cluster model, and the quark model (see also Morrison 2011). As Massimi explains:

> In high-energy physics, the nucleus is accurately represented by quantum chromodynamics as consisting of hadrons (protons and neutrons), each of which in turn consists of coloured quarks. But in the context of radioactivity and chemistry, the nucleus is best represented by the shell model, which represents the nucleons as occupying "shells"/ orbitals in some "magic numbers". If the phenomenon under study is nuclear fission, the liquid drop model best represents the nucleus and its behaviour. Finally, if the phenomenon of interest is the clustering

> of light nuclei (important in stellar nucleosynthesis), then the cluster model best represents the nucleus, with nucleons clustered in even and equal numbers (e.g. ^{4}He, ^{8}Be, ^{12}C). (Massimi 2022, 53)

8.2.1 MMI and Single-Model Preference

MMI looks to rule out in principle the possibility of arriving at a single correct model for some target system. If there is more than one model *M* that serves to represent a target system *S*, then $M_1 \ldots M_n$ cannot all be accurate representations of *S*. Hochstein notes that a consequence of "our dependence on such idealizing assumptions is that we are frequently left with collections of models that describe the same complex phenomenon in very different and often contradictory ways" (2023, 4).

Hochstein illustrates this with studies of the action potential of a neuron, among other examples. He notes that studies of this kind involve the use of a collection of highly idealized models and theories from many different domains, including: electrophysiology, chemistry, biology and physics. The result is a collection of different models, all of which are used to explain the action potential of a neuron but "with no way of merging them all into a single comprehensive representation" (2023, 5). Prima facie, therefore, the fact that scientific models use abstraction and idealization mitigates against the idea that it is possible to find a single best model for the explanation of a target system, or aspects of a target system. This result is built into the definition of MMI. This sentiment is echoed by Potochnik, who states:

> Combining representations of disparate influences into a single unified representation is oftentimes impossible, if only because of the limitations of our modeling techniques and computational powers. . . . In our complex world, visits to so many scientists engaged in different pursuits would not culminate in a full explanation of anything. At best, this would yield a hodgepodge of partially related representations, with no way to reconcile their incompatible assumptions and no way to combine their mathematical frameworks. (Potochnik 2017, 18)

This view is also endorsed by Dale, Dietrich, and Chemero, who make this claim specifically with respect to cognitive science: "The idea that one theory or framework can singly account for the vast complexity and variety of cognitive processes seems unlikely if not impossible" (2009, 1). MMI, therefore, appears to provide outright support for compatible pluralism but no support for competitive pluralism or any other account that strives for explanatory unification.

8.2.2 MMI and Model-Integration

MMI works especially well when multiple idealized models with inconsistent properties are used to explain different aspects of the same target system, since the aspects of *S* modeled by M_1 can be treated in relative isolation from the aspects of *S* modeled by M_2, and so on. In other words, the incompatibilities between the different model assumptions do not come into any kind of direct conflict. The upshot is that there will be no single model by which to represent aspects of *S*, but rather different models, $M_1, M_2 \ldots, M_n$, representing different aspects of the same target system. This would be an example of compatible pluralism.

Morrison (2011) provides an example of this kind of MMI. She considers multifaceted approaches to modeling fluid dynamics, especially the development of the boundary layer for fluids by Prandtl (1904). The idea of a boundary layer allows researchers to simplify the equations of fluid flow by dividing the flow field into two distinct areas: inside the boundary layer and outside the boundary layer. This facilitates a demarcation between two ways of modeling fluid dynamics. Inside the boundary layer, the viscosity of the fluid dominates, thereby creating the drag on an entity. Outside the boundary layer, the role of viscosity is neglected—it is idealized away by setting its value to zero. The importance of this development was that prior to Prandtl's model of fluid dynamics, the perfect fluid model was the dominant model. However, the perfect fluid model neglects viscosity. The problem with this is that even if cases involve only a very small viscosity, the difference between the ideal case and actual cases will be a high discrepancy. Prandtl's model allows one to divide conceptually a mathematical model into two sets of equations, for dealing with the inside and outside boundaries of the layer, respectively. As Morrison states:

> Prandtl's solution involved using different descriptions for different areas of the fluid that exhibit different properties. What this means is that we cannot have a single model of the fluid that represents the way it flows around a boundary. Despite the two different models they needn't be interpreted as inconsistent since they apply to different types of behaviour in different parts of the fluid. (Morrison 2011, 345)

8.3 Integrative Pluralism (in Biology) and Inferential Integration (in Cognition)

MMI works well for cases of compatible pluralism. I now want to consider two accounts of compatible pluralism in biology and cognitive science. I start with Mitchell's (2011) account of integrative pluralism in biology, and end with Hochstein's (2023) inferential account of integration in cognitive science.

8.3.1 Integrative Pluralism

Mitchell (2002) refers to Cartwright (1994) and Wimsatt (1987) to note the following: many scientific models of natural systems are literally false (the standard view of idealization described in chapter 2). For Cartwright, this means that many different models are required to explain the world, and that the models do so in a sort of patchwork way. One difference between Cartwright and Wimsatt is that Wimsatt takes it that second-generation models will perform better than first-generation models, even though they will still be idealized models. According to Mitchell: "Both Cartwright's and Wimsatt's analyses make sense of a compatible pluralism of scientific models" (2002, 63).

Mitchell (2002) considers a multiscale and complex phenomenon involving the behavior of an ant colony. According to integrative pluralism, if different models, say M_1 to M_3, are considered to be partial solutions to the same biological question, then one cannot integrate the various models into a single coherent model. Rather, because biological phenomena exhibit a multiplicity of causal pathways and historical contingency, the "type of integration that can occur in the application of models will itself be piecemeal and, to varying degrees, local. The result is that pluralism with respect to models can and should coexist with integration in the generation of explanations of complex and varied biological phenomena" (Mitchell 2002, 68). In other words, when one applies different idealized models, with different properties, to study a single target system, the various models will enable us to answer different questions about that target system, identifying different causal features of that target system and relevant constraints on explanation. However, we cannot get a single, unified model.

8.3.2 Inferential Integration

According to Hochstein (2023), it is widely assumed that the hippocampus and the entorhinal cortex play key roles in the construction of cognitive maps (i.e., spatial representations used, for example, by rats in navigating a maze). The physical vehicles of such cognitive maps are two different kinds of cells: grid cells and place cells. Each cell encodes different kinds of spatial information about the environment, which can be combined to form cognitive maps. Hochstein says:

> Now suppose we wish to provide an integrated account of this phenomenon. It turns out our ability to scientifically study this cannot be captured by any single model. This is because numerous contradictory models are required to describe what information exactly is being encoded by the different cells, how the mechanisms of the cells allow them to engage in this task, and how this information is being combined in the construction of cognitive maps. (Hochstein 2023, 10)

A reason why it is not possible, according to Hochstein, to produce a single model of cognitive maps is that different computational models are needed to describe the different information-processing tasks of the different kinds of cells and that these different computational models all involve idealizations and abstractions. According to Hochstein:

> This dependence means that computational models of neural circuits cannot be directly integrated with models that describe how the cell physically functions to engage in such computations, which in turn cannot be directly integrated with computational models that describe the way in which this information is combined in the construction of cognitive maps. (Hochstein 2023, 10)

Hochstein's proposal is that even if one cannot integrate these different models into a single, unified model, one can use idealized models with incompatible properties to draw coherent inferences across these models to target systems. For instance, drawing on the work of Garden et al. (2008), Hochstein proposes that physiological and structural models of grid cells, combined with other relevant background knowledge, place constraints on the development and understanding of computational models for information processing between cells. That is, one can use physiological and structural models of grid cells to make inferences about other models, such as computational models of information processing. What allows one to make such inferences in a coherent fashion is keeping track of ontological and metaphysical commitments. Hochstein (2023) appeals to the

Hardy-Weinberg model in population biology as an example. As we know, the Hardy-Weinberg model models real-world biological populations as being infinite. Yet, as Hochstein correctly notes, it would be a mistake to conclude from the success of such a model that population sizes must really be infinite. Scientists are not making inferences of that kind—this is consistent with the nonliteral view of idealization defended in part I of this book. Hochstein states:

> This is because if we look beyond what is explicitly stated in the model, to the implicit ontological and metaphysical commitments held by biologists, we find a commitment to the idea that real populations in nature are never infinite, and so it would be a mistake to draw that inference from such models. Instead, we use such models only to draw inferences about what traits would be locally optimal for an organism in a particular environment to have. Working scientists thus use their frequently unspoken or implicit ontological and metaphysical commitments not just as a guide for when and how to apply our models, but also for which inferences we are licensed to draw from them, and which we are not. Acknowledging and keeping track of these implicit commitments can be the key to understanding how incompatible models connect to one another. (Hochstein 2023, 9)

Hence, one can employ multiple different and idealized models with incompatible properties to explain one and the same phenomenon by noting that scientists use models to draw inferences about other models, thereby putting together a kind of inferential tapestry of a target system.

8.4 Compatible Pluralism and Species of Realism about Cognition

Integrative pluralism and inferential integration can be framed in terms of either deflationary realism or perspectival realism, or as a mix of both. Deflationary realism, as defined in Coelho Mollo (2021), is the view that models of cognition are partial and idealized and that many different models are therefore best suited in the study of cognition. Perspectival realism, as defined by Massimi (2022), claims that there is a perspectival element to any scientific theory. Scientific theories do not provide a "view from everywhere," but rather from a situated, historical perspective. More specifically, on Massimi's account of perspectival realism, the thesis subscribes to two specific assumptions:

> *The representationalist assumption.* Scientific models (partially) represent relevant aspects of a given target system *S*. A scientific model, or, better, the claims of

> knowledge delivered by a scientific model, are true (or approximately true) when the model provides a partial yet accurate representation of the target system. (Massimi 2022, 44)
>
> *The perspectivalist assumption.* Scientific models offer perspectival representations of relevant aspects of a given target system *S*. . . . Namely, scientific model M_1 perspectivally$_1$ represents *S* as *z*; but model M_2 perspectivally$_1$ represents *S* as *y* (where *z* and *y* are properties belonging to a "horizon of alternatives"). (Massimi 2022, 44)

Chirimuuta (2018) focuses on providing a computational model of vision. In doing so, she starts from a perspective on Marr's (1982) three different levels of explanation. Marr distinguished between the computational question, the representational or algorithmic question, and the implementational question. The first task for a theory of vision, according to the Marrian model, is to provide an abstract description of vision, where vision is taken to be a specific kind of computational problem. The second task is then to show how this computational problem can be solved by developing algorithms for transforming incoming visual information (e.g., turning the 2D pixel array at the retina into a 3D model of objects in the natural world). The final step tackles the implementation question—namely, how is it possible for a biological system to realize the algorithm to solve the computational problem?

We can leave the details of Chirimuuta's (2018) account of vision to the side, since here I want to focus on compatible pluralism in her account. First, to understand the relationship between the three levels of the Marrian framework, Chirimuuta contends that it is best to import the notion of 'perspectivism' from philosophy of science, where theoretical frameworks such as Newton's laws and Schröndinger's equations are taken not as objective windows on the world but rather as "perspectives" on the world. Chirimuuta (2018, 400) quotes Giere:

> General principles by themselves make no claims about the world, but more specific models constructed in accordance with the principles can be used to make claims about specific aspects of the world. And these claims can be tested against various instrumental perspectives. Nevertheless, all theoretical claims remain perspectival in that they apply only to aspects of the world and then, in part because they apply only to some aspects of the world, never with complete precision. (Giere 2006, 15)

The specific proposal is that we should treat each of the three Marrian explanatory levels as perspectives that scientists can use to model the visual system. Chirimuuta (2018) applies Bayesian models to the computational problem in vision science, partly because such models provide theoretically optimal strategies for dealing with uncertainty. According to Chirimuuta, the "mathematical theory of decision making gives a priori reason to think that Bayesian processing is relevant to biology, but it is just one perspective on a complex biological system" (2018, 405). Chirimuuta (2018) quotes Griffiths et al. in confirmation of this point:

> Different theoretical frameworks, such as Bayesian modeling, connectionism, and production systems, have different insights to offer about human cognition, distributed across different levels of analysis. A connectionist model and a Bayesian model of the same phenomenon can both provide valuable information—one about how the brain might solve a problem, the other about why this solution makes sense—and both could well be valid. (Griffiths et al. 2012, 421)

Each perspective is an idealized and partial model, and each model is applied to answer a different question in Marr's explanatory framework. As a result, Chirimuuta concludes that "we should be perspectival realists . . . and pluralists when interpreting Bayesian and other families of models in vision science" (2018, 404).

8.5 Inconsistent-Friendly Pluralism and Inconsistent Models

Compatible pluralism looks close to unavoidable given the accounts described in sections 8.3 and 8.4. MMI supports it. Does this follow? Not necessarily.

A central motivation for pluralism is that a single, unified, and complete model of some target system requires multiple different and idealized models; and because multiple different and idealized models of a target system will often have different and incompatible features, they cannot be integrated without loss of content (see, e.g., Kellert, Longino, and Waters 2006). Let us also assume that a central function of a scientific model is to represent a target system. Following Massimi (2022), let us further assume that by saying that the function of a scientific model is to represent a target system we take this to mean that a model accurately, or partially accurately, represents a target system. What we have here is a situation where pluralism can be said to be "inconsistency-friendly" (Massimi 2022, 54).

As far as I can tell, inconsistency-friendly pluralism describes deflationary realism, as defined by Coelho Mollo (2021). Deflationary realism is the view that scientific models are idealized and partial models of target systems, and that such idealized and partial models are well-suited to explaining certain patterns of behavior. According to Coelho Mollo:

> Given the complexity that marks cognitive systems and their interactions with their material and social environments, it is inevitable that different research aims and methods lead to a diversity of partial models, each bringing into focus one or a few explanatorily relevant factors, contributing a part of the story of how cognition and intelligent behavior are brought about. (Coelho Mollo 2021, 1064)

Here is the inconsistency-friendly position defended by Coelho Mollo:

> Embodied and enactive approaches to cognition offer distinct models, in light of often different explanatory aims. Representationalism, once seen under the light that I recommend, is not incompatible with such approaches, which tend to downsize or eliminate appeal to representations. Rather, we have a rich variety of partial models, embodying different idealizing assumptions and theoretical abstractions, all of which can be legitimately explanatory. Representationalism and antirepresentationalism are thus not competitors, but candidates for pluralistic integration. (Coelho Mollo 2021, 1064–1065)

Inconsistency-friendly pluralism is denied by both Mitchell (2002) and Hochstein (2023). For example, Hochstein warns us against the problem of using different models, with incompatible metaphysical commitments, to draw coherent inferences about one and the same target system. I agree. If one ignores what might be substantially different metaphysics across models, making use of both models to draw compatible or coherent inferences about one and the same target system will turn out to be problematic. In my view, inconsistency-friendly pluralism is not plausible. It is theoretically misleading. It is unproblematic to say that both representational and nonrepresentational models of cognition are partial and idealizing models. However, when taken at face value, these models ascribe different and conflicting properties to one and the same target system.

The problem with inconsistency-friendly pluralism is that it runs head first into the PIM. The PIM is "that different models may deliver different, sometimes incompatible, or even inconsistent images for the same phenomenon" (Massimi 2022, 53). The PIM poses a problem for scientific realism, including deflationary realism. Specifically, if a scientific model M_1 represents S as y (e.g., representational) and model M_2 represents S as z (e.g., nonrepresentational), then there is a significant tension between the two models; namely, M_1 and M_2 are representing S as having mutually inconsistent properties. However, realism is not compatible with the existence of two or more mutually inconsistent representations of one and the same target system. Therefore, inconsistent-friendly pluralism is not friendly to scientific realism.

The conjunction of MMI and the PIM is, however, based on a fallacy: the fallacy of adopting literalism about scientific discourse. It is a fallacy to move from the observation that multiple models are required to fully understand a target system to the additional claim that each model should be taken to literally represent that target system. This way of problematizing the PIM can be seen as friendly to perspectival realism. To see this, consider how Massimi (2019, 2022) seeks to overcome the PIM. Massimi (2022) does so by exposing two hidden premises that are tacitly linked to the representationalist assumption of perspectival realism (introduced in section 8.4). The first is the representing-as-mapping assumption:

> *Representing-as-mapping.* The true model is the one that offers an accurate, partial, *de re* representation of relevant essential features of the target system. Offering an accurate, partial, *de re* representation means to establish a one-to-one mapping between relevant (partial) features of the model and relevant (partial)—actual or fictional—states of affairs about the target system. (Massimi 2022, 66–67)

The second hidden premise is the truth-by-truthmakers assumption, according to which:

> *Truth-by-truthmakers.* States of affairs ascribe essential properties to particulars and, as such, they act as ontological grounds that make the knowledge claims afforded by the model (approximately) true. (Massimi 2022, 67)

Massimi's (2022) central claim is that only if these two additional premises are accepted, is the PIM a problem. Since perspectival realism is not committed to these two additional premises or assumptions, the PIM is no threat to perspectival realism. In chapters 3 and 4, I followed Rowbottom (2022) in arguing that scientific models can be described in the form of the sentence *S*, where *S* is = "p and p_1 . . . and p_n and *J*." I also argued that "p and p_1 . . . and p_n" are type (b) claims, where type (b) claims are nonidealizing, corresponding claims from a model to a target system. *J* is an idealization, which should be treated nonliterally, and exemplifies a type (a*) claim, where type (a*) claims are idealizing, noncorresponding claims (claims that do not have a representational function across the model-world relation). Hence, idealizing posits, which do not have a representational function from a model to a target system, do not align with the representing-as-mapping assumption as stated above. However, type (b) claims conform with the truth-by-truthmakers assumption, which separates a version of scientific realism based on nonliteralism about scientific discourse from perspectival realism.

8.6 Consistency-Friendly Pluralism: Does It Follow?

With the PIM arguably out of the way, the question now is whether MMI implies consistent-friendly, or compatible pluralism? To see why this is the case, it will be useful to further unpack perspectival realism, as it is formulated by Massimi (2022), and used by Chirimuuta (2018) in cognitive science. A central tenet of perspectival realism is the perspectivalist assumption, introduced in 8.4. It states that scientific models are "perspectival representations of relevant aspects of a given target system *S*. . . . Namely, scientific model M_1 perspectivally$_1$ represents *S* as *z*; but model M_2 perspectivally$_1$ represents *S* as *y* (where *z* and *y* are properties belonging to a 'horizon of alternatives')" (Massimi 2022, 44). What does it mean for *z* and *y* properties to belong to a "horizon of alternatives"? One way to capture the meaning of this phrase is through the familiar notion of representing-as (see, e.g., Elgin 2017; and chapter 3 in this book, where I framed this in terms of representation of *p* and *p*-representation). Massimi describes this view as follows:

> To classify a representation (in art or in science) as perspectival is often understood as implying that its *content* is *represented-as* *y* (rather than as *z*, or as *k*, or as *j*) *from* that specific vantage point. And *y*, *z*, *k*, or *j* are all attributes or properties belonging to the "horizon of alternatives" that the representation (*qua representation-as*) ascribes to the target. (Massimi 2022, 39–40; emphasis in original)

If we view this through the lens of MMI, we get the following picture: different models will represent the properties of a target system differently, such as having *z*-properties, as having *k*-properties, as having *j*-properties, and so on. As Massimi states: "Crucially, these attributes or properties may well be incompatible with one another, and this incompatibility is usually accepted as unproblematic in artistic representations, where an element of interpretation is inevitably present" (2022, 40). This is not how Massimi conceives of the "horizon of alternatives." In fact, if taken literally, it leads MMI to the PIM. Massimi makes this clear when she considers the representational content of perspectival realism:

> The representational *content* is not itself perspectival in that it is not affected by the vantage point *from* which the representation takes place: it is not an instance of *representing-as* understood as ascribing alternative and incompatible attributes or properties. Van Fraassen's [Fraassen 1980] hallmarks still apply, but the 'horizon of alternatives' does not capture rival or incompatible property ascriptions to the same target system. *It instead allows for a plurality of lines of inquiries and inferences about the target system.* (Massimi 2022, 41; last line of emphasis added)

This is what leads Chirimuuta (2018) to propose perspectival realism in the context of computational theories of vision. Perspectival realism is clearly a form of consistency-friendly pluralism, in the sense that there is a horizon of different perspectival alternatives by which one can make inquiries about a target system, where none of these different perspectival alternatives ascribe inconsistent properties to the same target system.

Chirimuuta (2018) seeks to justify a form of explanatory pluralism, what I have called consistency-friendly pluralism, by reference to the following claim by Griffiths et al. (2012) in a discussion of Bayesian models in computational neuroscience. They state: "Different theoretical frameworks, such as Bayesian modeling, connectionism, and production systems, have different insights to offer about human cognition, distributed across different levels of analysis" (2012, 421).

For Bayesian modeling, M_1, and connectionism, M_2, to provide genuinely different horizons of alternative perspectives, facilitating different lines of inquiry about the same target system, M_1 and M_2 must offer different sets of properties to allow for surrogative reasoning about one and the same target system. If M_1 and M_2 have the same properties, they will not offer different perspectives on the same target system, or the same aspects of a target system. That is, for M_1 and M_2 to enable surrogative reasoning about a target system, and for this to make a difference to our understanding of the target system under investigation, M_1 and M_2 had better have different properties. In other words, the variously different perspectives on a target system will allow for a plurality of productive lines of inquiries or inferences about a target system only if they do not share many of the same properties (see Hochstein 2023 for discussion). One way to understand this would be for M_1 and M_2 to provide us with multiple different and incompatible ways of perspectivally representing a target system. However, this would lead us back to the PIM. A different option would be to say that Bayesian modeling, M_1, captures relevant insight of the computational level proposed by Marr

(1982), while connectionism, M_2, provides us with ways of inquiring about aspects concerning the implementational level of the Marrian three-level framework. This is the line taken by Chirimuuta (2018). For the latter to succeed, it minimally requires that M_1 and M_2 are compatible. They are. The problem is that they are too compatible for M_1 and M_2 to function as different models of one and the same target system. As McClelland makes clear:

> Connectionist units that use the logistic or softmax activation functions can exactly compute Bayesian posterior probabilities when the bias terms and connection weights affecting such units are set to the logarithms of appropriate probabilistic quantities. Bayesian concepts such the prior, likelihood, (joint and marginal) posterior, probability matching and maximizing, and calculating vs. sampling from the posterior are all reviewed and linked to neural network computations. Probabilistic and neural network models are explicitly linked to the concept of a probabilistic generative model that describes the relationship between the underlying target of perception (e.g., the word intended by a speaker or other source of sensory stimuli) and the sensory input that reaches the perceiver for use in inferring the underlying target. (McClelland 2013, 1)

Bayesian modeling and connectionism can therefore be seen as part of one and the same computational modeling framework for understanding cognition. It is worth highlighting a slightly different point, starting with Marr's hierarchy of explanatory levels:

Level 1—the computational level

Level 2—the algorithmical level

Level 3—the implementational level.

As mentioned above, in Marr's (1982) theory, there are three different levels of analysis from which one can tackle questions about cognition, where the end goal would be to provide a multi-level or hierarchical explanation of cognition. Crucially, for Marr, these three levels can be viewed as relatively independent of one another. For example, one could seek to model the computational level in Bayesian terms, while investigating the implementational level through more neurally realistic connectionist models. This would allow M_1 and M_2, to provide genuinely different horizons of alternative perspectives, facilitating different lines of inquiry about the same target system, across relatively independent scales or levels of analysis.

One could arguably defend pluralism on that ground. But this will only work if these different explanatory levels can be treated as relatively independent levels of analysis, from which to tease out alternative perspectives on one and the same target system. In section 8.2, I highlighted the distinction between proximate questions and ultimate questions. Many take these questions to be independent of one another, and that those seeking answers to these types of questions work in different fields of research. If correct, this is a genuine case of compatible pluralism, since there is no tension between the questions under consideration—they can be given relative independent analysis. However, I also highlighted that the idea that these questions can be answered independently of one another is being questioned, and for good reasons. The same can be said of Marr's three-tier framework. Specifically, one might take the differences between computation, algorithm, and physical implementations as relatively insignificant. According to Sun:

> First, the differences among them are usually small, subtle, and graded, compared with the differences among phenomena to be modeled. Second, these different computational constructs are in reality closely entangled (especially in the psychological and the biological realm): one cannot specify algorithms without at least some considerations of possible implementations, and what is to be considered "computation" (i.e., what can be computed) relies on algorithms, including the issue of algorithmic complexity, and so on. Therefore, one often has to consider them together. (Sun 2023, 11)

However, once one starts to consider these different levels together, it is likely that there will turn out to be hardly any differences between M_1 and M_2, as highlighted by McClelland (2013) above.

At this stage, we can return to a discussion we had in chapter 7, where I suggested that one of the reasons why the mechanistic theory of computation idealizes is that it asserts that all computational systems are mechanistic systems. In computational neuroscience, one often finds examples of computational explanations that abstract away from biophysical details to examine certain kinds of why-questions about nervous systems. I also argued that one might treat the difference between computational and mechanistic explanation as a shallow difference. By "shallow difference," I mean that they approach the same kind of question about target systems from different routes. There is the computational high road, where one might start by considering questions of why systems exhibit the self-organizing dynamics that they do in order to derive general principles that seek to address these dynamics. I provided the free energy principle (FEP) as an example of the computational high road. The FEP is an efficient coding model. The FEP starts by asking what organisms must do to survive, and proceeds to formulate a set of principles by which it becomes possible to model how organisms survive. I also provided the FEP as an example of the mechanistic low road, which allows one to illustrate how organisms minimize entropy. Independently of which road one travels, one arrives at the view that organisms can be modeled as minimizing a variational (free energy) bound on entropy. It might be tempting to view this as an example of compatible pluralism. However, there is no pluralism here—just as there is no pluralism in the relationship between Bayesian models and connectionism, if the latter involves complex feed forward and backward connections (McClelland 2013).

8.7 Conclusion

In this chapter, I have looked at various reasons for thinking that explanatory pluralism is the way to go when having to explain complex phenomena. It did so by examining a well-known modeling practice, which Weisberg (2007, 2013) calls MMI. I showed that when model-based approaches to science, including the cognitive sciences, are viewed through the lens of MMI, it becomes tempting to conclude that any search for a single, unifying model of a target system is most likely to be unsuccessful. If this is the case, then we will often, if not always, have to use multiple different and idealized models when seeking to explain a target system. Pluralism, I then argued, comes in different forms, either as inconsistency-friendly pluralism or as more consistency-friendly versions of pluralisms. The former are unstable, as they marry MMI with the PIM. The latter, I argued, are pluralistic in appearance only—at least for the versions considered in this chapter. From here I went on to suggest that once the PIM is diffused, one can allow for pluralism to exist and start the process toward explanatory unification.

9 Idealization and a Grand Unifying Theory of the Brain (Case Study)

Overview: My intention in this chapter is to integrate all the many threads covered in this book into a single, coherent view of scientific realism by focusing on a particular case study in the form of the "free energy principle" (FEP) in theoretical neuroscience.

The FEP is known as a "grand unifying theory" (GUT). A GUT seeks to explain everything about the form and function of the brain by appealing to a small number of principles. GUTs are not found only in theoretical neuroscience. They are a key research goal in physics, starting with the pioneering work of Maxwell (1865). The FEP aims to provide a general theory unifying life and mind formulated entirely from mathematical principles in physics (Friston 2013; Hohwy 2020). The FEP is often said to be entirely scale-free, applying to any dynamical system able to maintain its organization despite tendencies toward disorder: from chemotaxis in cells, neuronal signaling in brains, tropism in plants, synchronized singing in birds, decision-making and planning in humans, and niche construction by human and nonhuman animals (Friston 2010; Hesp et al. 2019; Kirchhoff 2015; Kirchhoff 2018b; Kirchhoff and Kiverstein 2019; Kirchhoff et al. 2018; Palacios et al. 2020; Ramstead et al. 2021).

The FEP is controversial. However, the FEP has seen an explosion of interest across the sciences and philosophy in the last decade or so. Moreover, the FEP lends itself as a perfect case study by which to unify all the different threads covered in this book. Specifically:

- The FEP is an idealized, abstract and mathematical model (chapter 1).
- The FEP is not literally true of its intended target domain (chapters 2–3).
- The FEP is consistent with a version of scientific realism based on the nonliteral view of idealization (chapters 3–4).

- The FEP does not lend support to the RTM (chapters 5).
- The FEP is a computational model yet lends no support to the CTC (chapters 6–7).
- The FEP represents an extreme form of explanatory unification (chapter 8).

My aim in this chapter is not to convince anyone that the FEP can deliver on its promises to be a genuine GUT (for discussion, see Mann, Pain, and Kirchhoff 2022). Rather, my aim is to show how the FEP can serve to unify the claims listed above.

9.1 Introduction

In chapter 1, I reviewed three different kinds of modeling practices: targetless modeling, generalized modeling, and target-directed modeling. I argued that in targetless modeling, questions about realism versus antirealism do not arise. Targetless modeling, I claimed, is a safe space for realists and antirealists alike. The real tension in the discussion over scientific realism gets out of the starting blocks once questions are asked across the model-world relation. I called this the "map problem."

In this chapter, I consider this issue afresh in relation to the FEP. "Free energy" is a mathematical term. It is defined by differences between probability distributions, where free energy quantifies the difference (or, formally, divergence) between predictions about sensory signals and the actual sensory signals received. The FEP states that any process that occurs in the brain can be modeled starting from a single principle: minimizing free energy, where free energy functions as an upper bound on entropy. Entropy is the average self-information or surprise (where "surprise" is the negative log-probability of some outcome). The basic idea is that, by minimizing free energy, a self-organizing system can minimize surprising encounters with the world, where "the world" can be both the internal, bodily world, and the external, extra-bodily world. A standard example is given by Friston: "A fish out of water would be in a surprising state (both emotionally and mathematically). A fish that frequently forsook water would have high entropy" (2010, 127). Surprise is therefore species-dependent. What is surprising for a fish need not be surprising for you, since being on land for humans is being in a state of low entropy, all things being equal.

According to the FEP, organisms have two ways of minimizing free energy, which can be achieved by optimizing two functions: a variational free energy function—measuring the fit between a model of sensations and actual sensations—and an expected free energy function—scoring possible action policies by which to remain in or reach preferred states of being (Da Costa et al. 2021; Friston 2013; Ramstead, Badcock, and Friston 2019). The former option involves updating prior beliefs about the world based on sensory data. Here the internal machinery is in many ways controlled by the experiential world. The latter option expresses the idea that free energy can be minimized by controlling the experiential world itself—by changing the causes of sensory input. Grace puts this nicely with the following illustration:

> Imagine a bird that has grown accustomed to flying around a certain forest; it can predict which trees will be good for nest building, where the best food is found and so on. One day it flies a little beyond its normal range and finds itself in a city. Experiencing tall buildings and traffic for the first time, its ability to predict almost anything about the world around it is low. This big discrepancy between prediction and experience means free energy is high. To bring its free energy back down, the bird could stick around and hope its sensory systems adapt to be able to predict the features of city life. Or it could simply fly back to the forest it came from. The presence of this second option—choosing actions that result in predictable sensory experiences—is what makes the free energy principle a candidate GUT of the brain. Rather than just explaining features of sensory processing, this principle can encompass decisions about behaviour as well. (Grace 2021, 332)

The FEP has enjoyed a huge amount of attention in recent years, across theoretical neuroscience and theoretical biology and from philosophers of cognitive science. The last five years, in particular, have seen an interesting increase in research defending instrumentalism about FEP models. Instrumentalists about FEP models rely on the following kind of argument to supports their antirealist leanings:

1. Scientific models introduce distortions into their representations of target systems via idealization and approximation.
2. Scientific realism requires that models provide descriptions of target phenomena that must be taken literally; either true or false (literalism).
3. Scientific models are not true and accurate representations of their targets.
4. Therefore, scientific realism about models is false (presupposition: veritism).

This kind of argument rests on three propositions that I have discussed throughout this book: (1) scientific discourse must be taken at face value, as truth-conditioned descriptions of its intended target domain (literalism); (2) truth is necessary for epistemic acceptability (veritism); and (3) given the pervasive use of idealization in scientific modeling, much of what science produces is false, if taken literally. The instrumentalist view on the FEP endorses (1): scientific realism about the FEP requires that models built based on the FEP must be taken literally; they are either true or false. However, (3) is entailed by (1). Therefore, if the FEP does not yield literally true models of the structure and function of the brain, realism about the FEP must be denied. Consider, for example, what the following authors say about the FEP:

- Ramstead, Friston, and Hipólito (2020, 4): "Instrumentalist accounts in the philosophy of science suggest that scientific models are useful fictions: they are not literally true, but 'true enough,' or good enough to make useful predictions about, and act upon, the world."

The instrumentalist argument is misguided. First, if scientific models are fictions, then they are not candidates from being true; not even "true enough." Second, the instrumentalist argument as construed above rests on a fallacy: the literalist fallacy (Kirchhoff, Kiverstein, and Robertson forthcoming). For variations of this fallacy, see chapter 4. Here it takes the following form. The literalist fallacy, in the discussion about the FEP, is the fallacy of inferring the truth of instrumentalism based on the claim that the properties of FEP models do not literally map onto target systems. The literalist fallacy is committed by anyone that infers that scientific realism cannot be defended in cases where scientific models make use of idealizations—whether temporary or not. What I argue in this chapter is that denying literalism about some scientific discourse opens up a space for defending scientific realism about the FEP. I rely on the arguments developed in chapters 3–8 to justify this claim. The FEP is a model-based approach to theoretical neuroscience. Like the HH model, or the Lotka-Volterra equations of predator-prey dynamics, the FEP is a mathematical model that trades in both abstraction and idealization. However, the idea that scientific models involve idealizations is not a threat to scientific realism precisely because idealizations should be treated nonliterally. Therefore, idealization is not a problem for scientific realism about the FEP.

I shall proceed as follows. In section 9.2, I focus on the relationship between the FEP and idealization. In section 9.3, I consider arguments in favor of taking variational free energy to be a useful fiction. In section 9.4, I discuss the role of the Markov blanket formalism underwriting the FEP as a mathematical model. In section 9.5, I discuss the status of generative models in the FEP. In section 9.6, I provide a set of arguments for why the FEP does not support realism about representationalism. I conclude in section 9.7.

9.2 Idealization and the Free Energy Principle

Computational neuroscience proceeds, in large part, by building models for the purpose of evaluating their biological plausibility. As with the FEP, such models make use of techniques from machine learning. Hence, a few words are needed about this relationship—between machine learning and neuroscience. First, machine learning techniques allow scientists to process and analyze large data sets from experiments and measurements (Pain, Kirchhoff, and Stephen Mann 2022). The variational techniques imported into neuroscience from machine learning are used to solve problems with computational intractability (Dayan and Abbott 2001). This was one of the main functions of idealization introduced in chapters 2–3. The FEP models the brain as facing a problem akin to the experimental scientist: the scientist and the brain must use the data they have to draw conclusions about events they cannot observe directly. Mann, Pain, and Kirchhoff (2022) call this "the general relationship" between machine learning and neuroscience, where machine learning techniques are used to solve general computational problems. When viewed as a general relationship between machine learning and neuroscience, the relationship between these two domains of research is like the relationship between machine learning and any other branch of science where techniques from machine learning are introduced to compute large datasets. There is also a "specific relationship" between machine learning and neuroscience. The specific relationship between machine learning and neuroscience is when questions are asked about the relation between a specific model or a specific mathematical model and the actual target system. The specific relationship is the same as the map problem: What is the relationship between a model and its intended target system in the world? With regards to the FEP, it is the special relationship I am interested in. Consider figure 9.1 from Friston (2010).

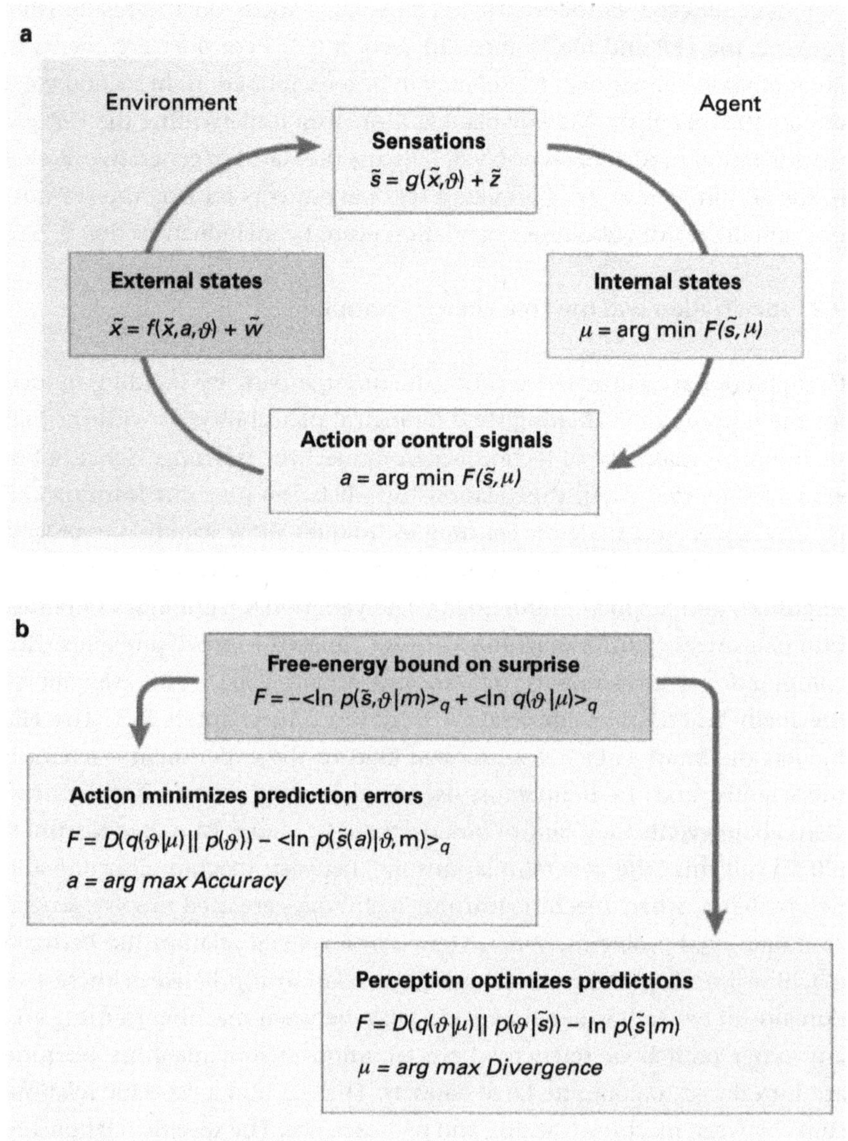

Figure 9.1
The free energy principle (in Friston 2010, 128). Used with permission from the Wellcome Trust.

Part (a) expresses the dependencies between the different quantities that define free energy, which includes internal states of the brain $\mu(t)$ and the quantities that describe how the internal states are related to external states: sensory signals, including their respective dynamics $s(t) = [s, s', s'' \ldots]^T$ in addition to action $a(t)$. The external states (or environmental states) are described in terms of equations of motion, which, from the point of view of the FEP, specify the trajectory of hidden states (they are called hidden, since the system in question does not have access to these states). The hidden causes are formulated as $\vartheta \supset \{x, \theta, \gamma\}$. They are the causes of sensory input, which comprise hidden states $x(t)$, parameters θ and precision γ, which conjointly control the amplitude of random fluctuations $z(t)$ and $w(t)$. Crucially, it is the internal states and active states that minimize the free energy $F(s, \mu)$, which is a function of sensory input and a probabilistic distribution over its causes $q(\vartheta \mid \mu)$. This probabilistic distribution is also called the recognition density. The free energy formalism depends on two probabilistic densities: the recognition density and a density that generates sensory input and their possible causes, $p(s, \vartheta \mid m)$, which refers to a probabilistic generative model (m). I will leave the formalism behind for part (b), and simply note that it provides two expressions for how free energy can be minimized. Active states (or action) can reduce free energy only by increasing accuracy (that is, changing to causes of sensory input to fit with expectations about worldly input). Alternatively, free energy can be minimized by optimizing the fit or divergence between the posterior generative and approximate recognition densities. The former is often called *action*, and the latter *perception*, bringing the mathematical formalism of the FEP into contact with biology and cognition.

What does this tell us about the specific relation between the FEP qua a mathematical model and its target system (the brain)? Klein (2018) has suggested that the FEP is best thought of as Galilean idealization. Klein describes the FEP as being

> literally false, but with some understanding gained via over-simplification. Such idealizations can often be elaborated to be true of particular systems, and those elaborated models—which often look very different from the original model—can have considerable explanatory power. But if this is the case, then it is worth keeping in mind that FEP is a starting point from which one might develop explanations, and that its defence would ultimately rest on the empirical adequacy of detailed models which spring from it. Simplicity does not count in its favour, for

> FEP is simple in the way that friction-free planes and infinite populations of bunnies are simple: that is, a deliberate simplification, which buys scientific fruitfulness at the cost of literal truth. (Klein 2018, 2253–2254)

There are several things to note here. First, Galilean idealization is one type of idealization. It is defined as a simplification or distortion of something into a model to turn a computationally intractable problem into one that is tractable (McMullin 1985). For the FEP, the variational notations for the free energy functions would be candidates for being Galilean idealizations. However, Galilean idealization is an instance of the "falsehood view of idealization" (FvI), which I rejected in chapters 3–4. One can use the Kullback-Leibler (KL) divergence to measure the distance between the posterior generative and approximate recognition densities without thereby saying that it is literally true that the nervous system encodes the KL divergence (part [b]—perception optimizes predictions). This follows from the discussion in chapters 6 and 7. In machine learning, the KL divergence is a way of measuring the dissimilarity of two probability distributions, p and q. It provides a way of quantifying the relative entropy between two distributions (Murphy 2012, 58). The KL divergence is a modeling technique that may assist with formulating specific assumptions about the structure and function of specific systems, with the aim of generating testable empirical claims. As with the HH model, it is an idealized formalism. However, as I have argued, one need not view idealized posits as either true or false of target systems. Klein's (2018) assessment of the FEP is problematic because he is working with the FvI in thinking about idealization.

Second, Weisberg (2007) draws our attention to the fact Galilean idealization is consistent with scientific realism. He says of this kind of idealized model that it is the most straightforward type of idealized model compatible with scientific realism. Accordingly:

> the Galilean idealizer does aim to give complete, non-distorted, perfectly accurate representations. In order to accommodate the possibility of Galilean idealisation, scientific realists need to understand that achieving accurate representations of complex phenomena is an ongoing process. Even when the short term practice involves the willful introduction of distortion, the long-term aim can still be to give an accurate representation of what reality is really like. Thus scientific realism is perfectly compatible with Galilean idealisation, if the realist aim is understood to be long term or ultimate. (Weisberg 2007, 657)

According to Weisberg, Galilean idealizations are, in principle, temporary or dispensable (see also chapters 2, 4). As I see it, the only problem with

Galilean idealizations is that they are deemed falsehoods. De-idealization is not a problem. However, idealized posits in the FEP look to be indispensable. Within the FEP, free energy is formulated mathematically as an upper bound on surprise, since surprise is widely recognized as being a computationally intractable quantity. The main reason for this has to do with having to summarize over the joint probability distribution between causes of sensory input and sensory input. However, computing the surprise associated with sensory observations would require complete access to all of the possible external states that could potentially cause sensory input (Friston 2010). Such a computation is intractable. This is why variational free energy is relevant. It is associated with a proxy for the quantity of surprise elicited by sensory stimuli, and defined as a *functional*: it is a function of the function of sensory and internal states. It can be employed to reduce surprise precisely because it is defined as being equal to or greater than surprise. For a formal proof put in terms of *Jensen's inequality*, see Kirchhoff et al. (2018, appendix A; and chapter 10 below). Therefore, contrary to Klein (2018), I conclude that the FEP is not an example of Galilean idealization—independently of interpreted literally or nonliterally.

Third, even if we grant that the FEP is an idealization, it does not follow that the FEP is literally false, in the sense of being completely false. As I argued in chapter 4, scientific models consist of two types of claims: (a*) idealizing, noncorresponding claims; and (b) nonidealizing, corresponding claims. Type (b) claims are candidates for a literal treatment. Type (a*) claims, however, should be treated nonliterally. For instance, in his work on projectile motion or the diurnal rotation of the Earth, Galileo made use of models with type (a) and type (b) claims. Some descriptions of entities should be taken as literally true, such as claims about projectiles, the Earth, and so on. Other descriptions are better constructed as idealizing and noncorresponding claims, such as claims involving frictionless planes and so on. Similarly, the FEP puts forth nonidealizing, corresponding claims (claims about neurons, reflex arcs, the hierarchical organization of the nervous system, and so on). It also puts forth idealizing, noncorresponding claims (such as the variational free energy notion). Treating type (a*) claims nonliterally allows us to say that type (a*) claims are not literally false with respect to target systems.

Finally, in chapter 2, I introduced two different views of idealization. One is the temporary view. An example of this is Galilean idealization, discussed above. The other is the permanent view, according to which idealizations are ineliminable for the practice of scientific modeling because of their epistemic benefits. There are at least three distinct versions of the permanent view. I will discuss two of them here.

According to Strevens, a good idealizing explanation is one that provides us with the explanatorily essential information, and it "is always better than its veridical counterpart" (2008, 318). Hence, for Strevens, since idealization serves to flag which parts of a system are non-difference-makers, idealization facilitates the identification of causal difference-makers. Idealization is, therefore, essential to scientific modeling. Is the FEP an instance of Strevens idealization? It is hard to see how it could be. Strevens idealization attributes explanatory relevance only to the accurately represented parts of a model, since it is those parts that are deemed to be causal difference-makers and it is those parts that overlap between the model and the target system.

Batterman and Rice (2014), in contrast, argue that Strevens idealization is problematic because it overlooks what makes many idealized models explanatory; their actual idealized posits. It is a virtue to have overlapping parts between a model and its intended target system. Yet, according to Batterman and Rice (2014), what is critical to the explanatory power of a model is its idealizations. They propose what is called the holistic view of idealization (elaborated in Rice 2018, 2019). According to Rice (2019), idealized models often fail to be decomposable into accurate and inaccurate parts, where the latter can be flagged as explanatorily irrelevant. They fail decomposability, since it is often the idealized posits themselves that are crucial for explanation. While Rice (2019) still adopts the FvI, Suárez (2010) goes one step further. Here is a slightly modified representation of the idea, which we can call the "inferential view of idealization":

> **Given B_1 & B_2 & [. . .]: If F_1 & F_2 & [. . .] then M_1 & M_2 & [. . .]**, where {B_1, B_2, etc.} are nonidealized background assumptions, {F_1, F_2, etc.} are idealized assumptions within the model, while {M_1, M_2, etc.} are measurable properties of the target systems.

My proposal is that this captures how idealization figures into the FEP. Let B_1 & B_2, & [. . .] represent the following nonidealized background assumptions. Living organisms must constantly modulate their relation to their environments (including other organisms). If they do not do this, they will likely die. For instance, organisms will continuously and repeatedly need to feed themselves, and so on. In doing so, organisms act in the world (thereby changing it) and receive sensory stimuli from it. Hence, living organisms can maintain their integrity only by exerting adaptive control across continuous cycles of action-perception loops (Parr, Pezzullo, and Friston 2022). Let F_1 & F_2 & [. . .] be the FEP formalism represented in figure 9.1 above. According to this view, the point of idealization is to provide us with a set of idealized conditionals, the function of which is to facilitate inferences from B_1 & B_2, & [. . .] to measurable consequences, M_1 & M_2 & [. . .]. This view of idealization is a natural description of the FEP. Indeed, Hohwy (2020) argues that the generalized form of the FEP should be treated as a regulatory principle guiding theory construction of process theories. A process theory, in this context, is about how surprise minimization models such as the FEP can be realized by neuronal (and other biophysical) processes in the brain and beyond (Parr, Pezzullo, and Friston 2022, viii). Treating idealizations in the FEP as part of the conditional antecedent from which to infer claims about testable consequences is part of Hohwy's (2020) view of the FEP (as far as I can tell); that it can be used to make inferences about the structure of specific systems for the purpose of yielding testable empirical claims (which would include computational, algorithmic, and implementational claims about neural activity). In this sense, the FEP can be used to discover mechanisms for surprise in biological systems (following from chapter 8).

9.3 Variational Free Energy and Fictionalism

In the context of FEP models, the variational free energy notation is a good example of a theoretical entity that can be given a concrete mathematical specification, although target systems do not literally reflect the mathematical structure of the variational free energy notation (from chapters 6–7). Yet, instrumentalists about the FEP conclude, on this basis, that because the free energy notation is not literally true of a target system it is simply a useful fiction. Let us see how they arrive at this conclusion.

Variational free energy is an information-theoretic construction used to compute how organisms can resist decay by positing variational free energy as an upper bound on surprise. In FEP models, for an organism to exist it must keep its states (e.g., homeostatic states) within certain bounds. Organisms must maintain a low conditional entropy over their internal states (Corcoran, Pezzulo, and Hohwy 2021). FEP models invoke an information-theoretic term to explain how this is achieved: surprise. This term allows a quantification of the improbability of some outcome (e.g., some sensory data). FEP models state that by minimizing the surprise associated with sensory states, an organism can keep the entropy of its states low (and effectively survive), since entropy converges with long-term surprise.

Friston says that the FEP is "a mathematical formulation of how adaptive systems (that is, biological agents, like animals or brains) resist a natural tendency to decay" (Friston 2010, 1). Crucially, according to Ramstead, Badcock, and Friston, "Any system that avoids surprising exchanges with the world (i.e., surprising sensory states) will look *as if* it is predicting, tracking, and minimising a quantity called variational free energy, on average and over time" (2019, 320; emphasis added). This suggests that what is implied is that "the system does not actually predict, infer, track or minimize a quantity called variational free energy, but it merely looks *as if* this is what it is doing. The probabilistic model merely tracks certain real statistical relations in the organism-environment system" (van Es 2021, 321). Since target systems do not literally reduce this quantity, these authors conclude that scientific realism about FEP models is problematic.

Variational free energy is an idealized notation in FEP models because it is not literally true that target systems minimize this quantity. This is true of all models that make use of idealizations. Batterman (2009) states that all scientific models involve idealization. So, if the argument in favor of instrumentalism is correct, scientific realism cannot be defended in any model-based scientific domain. The arguments in chapter 4 blocks this result.

The FEP raises important modeling issues and questions. For example, it can be used to "quantify and simulate self-evidencing [i.e., maximize model evidence]" (Friston 2019, 85). It prompts the question of whether "self-organization approximates Bayesian inference—or does Bayesian inference approximate self-organization?" (Friston 2019, 85). These are modeling questions that may lead to new discoveries about target systems. This is consistent with the view that idealized posits may form parts of an antecedent structure from which to draw inferences about the properties of target systems. I will come back to this issue. For now, I want to focus on the following: the status of variational free energy as an idealized entity has lent false credibility to instrumentalist claims about FEP models.

Here is how the as-if claim about FEP models leads some to affirm instrumentalism about the FEP. If taken literally, FEP models cannot be true. Alternatively, FEP models may not be understood literally. However, if FEP models should not be understood literally, then scientific realism about

FEP models is false. So it seems to those favoring instrumentalism about FEP models. This apparent dilemma for the scientific realist rests upon a misunderstanding of scientific realism. Instrumentalism about FEP models does not follow from the fact that FEP models do not provide descriptions of target systems that are literally true. Or, differently put, one cannot justify instrumentalism about FEP models because such models are idealized models. This much follows if one acknowledges that scientific models need not be literally true for them to provide approximately true explanations of systems in reality (chapter 4).

This raises an important question: What is the status of type (a*) claims in scientific models? Recall, type (a*) claims are idealizing, noncorresponding claims. Some proponents of the instrumentalist reading of the FEP say that the claim that organisms minimize free energy is a useful fiction—it is an "as if" story, not a literally true story. It is tempting to think that if a claim is not literally true of a target system, but is nevertheless explanatorily useful, then this claim must have the status of a fiction. It is this fictional status of variational free energy that motivates instrumentalism about FEP models.

It is possible to treat the variational free energy notation as akin to fictional constructs in literature. However, this does not entail a rejection of scientific realism. Godfrey-Smith (2009) defends this view about idealization in science, and he does so while assuming a kind of scientific realism. Godfrey-Smith (2009) notes three important properties of fictional entities in science. First, hypothetically, were such entities to exist, they would be entities located in space-time. They would allow for actual, not hypothetical, interventions. Second, fictional entities have investigative properties; they are the common property of a community of researchers—for example, the FEP community, which inherits many of its notations from other areas such as statistical physics and machine learning. This means they can be "investigated collaboratively [and] surprising properties might be uncovered by one investigator after being denied by another" (Godfrey-Smith 2009, 102). Finally, given their hypothetical *and* investigative properties, "their status, though not their role, . . . seem analogous to the fictions of literature" (102). The crucial point here is that by means of theorizing with fictional entities, one may learn about the world—with respect to both observable entities and unobservable structures. This is enough for scientific realism, and it is sufficient vis-à-vis models generated through the FEP:

it posits fictional entities in its modeling of target systems for scientists to learn about such systems.

Does the FEP qua its idealized status have this kind of investigative status? I think that the answer to this question is a positive one. According to Turner and Zandt, new Bayesian techniques using approximate methods are making it "possible to fit these . . . models to data. These techniques have even allowed simulation-based models to transition into neuroscience, where tests of cognitive theories can be biologically substantiated" (2018, 1). In the FEP literature, this sort of work is starting to emerge. FEP models are being used to explore how biological systems such as the brain can anticipate sensory observations. For example, Parr et al. (2019) investigate FEP models in active vision. Of the significance of this work, they say:

> In brief, we used magnetoencephalography in combination with eye-tracking to assess the neural correlates of a form of short-term memory during a dot cancellation task. Using dynamic causal modelling to quantify changes in effective connectivity, we found evidence that the coupling between the dorsal and ventral attention networks changed during the saccadic interrogation of a simple visual scene. Intuitively, this is consistent with the idea that these neuronal connections may encode beliefs about "what I would see if I looked there", and that this mapping is optimized as new data are obtained with each fixation. (Parr et al. 2019, 1)

Of course, one might still try to press the following point. There are models that cannot be understood literally, even if they are useful models; conversely, there are models that cannot be understood literally and the reason is that they are not explanatorily useful or fruitful. Are FEP models explanatorily useless? That is far from obvious. FEP models are being used to gain insight into a diverse range of cognitive phenomena: decision making under uncertainty (Friston, Samothrakis, and Montague 2012), psychopathology (Schwartenbeck et al. 2015), active scene construction (Mirza et al. 2016), and so on (see Da Costa et al., 2020 for a detailed overview of different applications of FEP models).

I do not think that idealizations are, necessarily, fictional in the specific sense of fictionalism. I will say more about this in chapter 10. Here I will sketch in further detail why fictionalism as such does not undercut any quest for scientific realism. Consider what Suárez (2010) calls the *incompatibility argument*:

- Premise 1: All scientific representation involves fiction, i.e., it is either fictional or fictive.

- Premise 2: A fiction's cognitive function in inquiry is independent of its truth-value (or degree of accuracy).
- Premise 3: If (1) and (2), then science does not aim at truth.
- *Therefore*: Science does not aim at truth: Scientific realism is false.

Premise 3 is arguably false. In what follows I do not ascribe what I am arguing to Suárez (2010). I am using the incompatibility argument to bring out what is one of the central messages of this book; namely, that one can accept premises 1 and 2, while denying premise 3 and thereby block the conclusion. One might accept that all scientific models involve fictions and one might also accept that the cognitive function of a fiction is independent of its truth-value. Premise 2 is a nonstandard view of fictions (or idealizations). The standard view is the truth-conditional account, which is a version of the falsehoods view of idealization (chapter 2). On the truth-conditional account of fiction, falsehood is a defining property of what it is for something to be fictional. Accordingly, for a posit in a model to function as a fiction it must be: (1) truth-apt—this is the claim that scientific discourse, even when involving fictions must be taken at face-value (literalism); (2) it must be false—which it is when viewed across the model-world relation; and (3) those working with scientific models of this kind must be aware of (1) and (2). One of the arguments of this book is that the truth-conditional account is problematic. If the function of an idealized posit is not to represent aspects of the world, it is not a falsehood with respect to that aspect of the world (chapter 3). Hence, the truth-conditional account is questionable. Premise 3 is false because one can deny the truth-conditional account without having to accept the conclusion. One reason is provided by Suárez (2010): fictions provide inferential shortcuts in models by which one can make testable inferences about the empirical world. This is consistent with scientific realism. Thus, the appeal to the fictional status of the FEP does not license any kind of direct claim concerning instrumentalism about the FEP; nor of any other kind of scientific model.

9.4 Markov Blankets

Bruineberg et al. (2021) argue that the use of the Markov blanket formalism in the literature on the FEP often conflates the model with the target system. The Markov blanket concept originates with Pearl (1988) in his work on probabilistic reasoning and Bayesian networks. Pearl introduced the notion of the Markov blanket to denote a set of topological properties specific to Bayesian nets. See figure 9.2.

The Markov blanket in figure 9.2 is the smallest set of nodes {2,3,4,6,7} that renders a target node {5} conditionally independent of all other nodes in the model {1}. The central point to note here is that the behavior of {5} will be predictable by knowing the nodes making up its Markov blanket. This means that any node external to the system in question—in this case, node {1}—will be uninformative vis-à-vis predicting the behavior of {5}. This means that once all the neighboring variables for {5} are known, knowing the state of {1} provides no additional information about the state of {5}. It is this kind of statistical neighborhood for {5} that Pearl called a Markov blanket. Take a very simple case of conditional independence. Observing it being cold could be explained by a window having been left open, or by excessively high air conditioning. On observing that the air conditioning is on full, the observation that it is cold now carries no information about whether the window is open (under the assumption that the two events are statistically independent). What this example highlights is that the coldness in the room and the window being open are conditionally independent, given that the air conditioning is on high. We can put this formally, noting that for any variable *A*, *A* is conditionally independent of *B*, given another variable *C* if and only if the probability of *A* given *C* and the probability of *B* given *C* can be written as $p(A|C)$ and $p(B|C)$. In other words, *A* (it being cold) is conditionally independent of *B* (the window

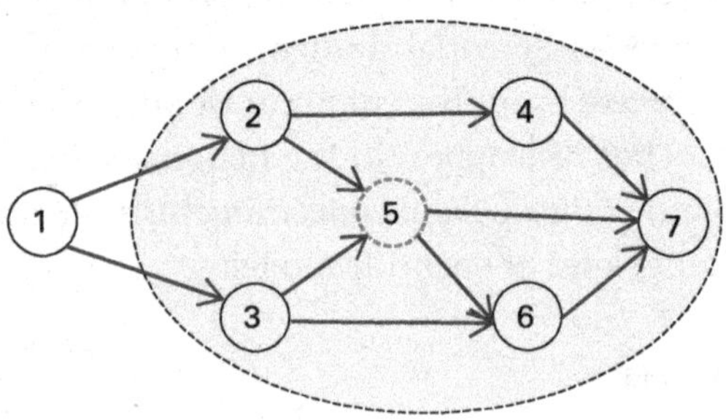

Figure 9.2
A schematic depiction of a Markov blanket with full conditionals (Kirchhoff et al. 2018).

being open) given *C* (high air conditioning) if, when *C* is known, knowing *A* would provide no additional information about *B* (Beal 2003).

Markov blankets can be used to model (in)dependencies between different variables, which allows for an approach to probabilistic reasoning under uncertain circumstances. Bruineberg et al. (2021) call Markov blankets of this kind *Pearl blankets*. Under the FEP, this formal model of identifying conditional independence between nodes in a network is interpreted in a specific way. That is, once a Markov blanket has been identified for a living system, the blanket states are modeled as active and sensory states, separating internal (organismic) states from external (environmental) states. Bruineberg et al. (2021) call Markov blankets of this kind *Friston blankets*. Unlike Pearl blankets, Friston blankets work to describe a real-world boundary demarcating an organism from its environment, and vice versa. See figure 9.3.

Figure 9.3 highlights the partitioning rule governing Markov blankets. Hidden states (Ψ) cause sensory states (S), which influence, but are not themselves influenced by, internal states (R), while internal states cause active states (A), which influence, but are not themselves influenced by, external states. Internal and external states can therefore be understood as influencing one another in a continuous and reciprocal fashion, given how sensory states couple (i.e., relate) external states to internal states, while active states couple internal states to external states.

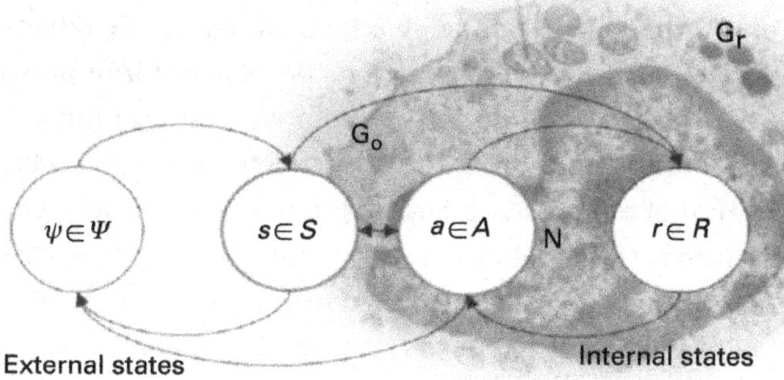

Figure 9.3
The partitioning rule governing the dependencies and independencies induced by a Markov blanket (Friston 2013). Used with permission from the Wellcome Trust.

According to Bruineberg et al. (2021), the use of Friston blankets to draw a distinction between organism and environment is not licensed by appeal to the mathematical structure of Pearl blanket. Hence, Bruineberg et al., finish their article by providing FEP researchers with a choice:

> The considerations presented in this paper leaves [*sic*] the FEP theorist with a choice. One can accept a rather technical and innocent conception of Markov blankets as an auxiliary formal concept that define[s] what nodes are relevant for variational inference [=Pearl blankets]. This conception is admittedly scientifically useful but has not yet led to any philosophically interesting conclusions about the nature of life or cognition. Alternatively, one can import a number of stronger metaphysical assumptions about the mathematical structure of reality to support a realist reading, where the blanket becomes a literal boundary between agents and their environment [=Friston blankets]. Such a strong realist reading cannot be justified by just "doing the maths", but rather needs to be independently argued for, and no such argument has yet been offered. (Bruineberg et al. 2021, 17)

The problem with this kind of argument is that accepting some important, or even fundamental, link between Markov blankets and organism-world boundaries does not entail accepting the further literalist claim, that the formal and mathematical structure of a Markov blanket (the model) literally is the boundary (the target system). Hence, Bruineberg et al. (2021) confuse what researchers working with Friston blankets need to be committed to.

Bruineberg et al. make a good deal out of the fact that the shift from Pearl blankets to Friston blankets requires an interpretive step. It requires interpreting Markov blanket states in terms of sensory and active states. But, any kind of interpretation is done by a theorist, rendering the identification of Markov blanket states arbitrary. The point about interpretation is fair. It is another thing entirely to infer that interpretation implies arbitrariness. Science is unlikely to progress in the absence of interpretation. First, any application of a model to a target system requires interpretation of the model for it to be a model of the target system. This is the case for any kind of model, whether it be concrete, mathematical, computational, or some other. Recall our example of the ship model in chapter 1: even a scale model of a ship must be interpreted as a model of a ship (the actual SS *Monterey*) for it to be meaningful for those using it. Interpretation is part of any modeling exercise in science and nothing about the need for interpretation implies any kind of antirealism about scientific theories (and models). Nor does the use of interpretation rule out scientific realism about the use of the Markov blanket formalism in the application of FEP models.

Model building involves interpreting structures in the world in a meaningful way. If this is on the right track, then science without any kind of interpretation, without any kind of explanatory or interest on behalf of researchers is a false idealization of any type of scientific method. Interpretation also serves a different role. Consider Nguyen and Frigg's (2022) discussion of what makes a model a model of a target system. One option is that the model looks like its target system—a small-scale ship model that looks like its target system (chapter 1). However, often the models used in science do not at all look like the target systems they are invoked to model. Nguyen and Frigg give the following examples:

> Imaginary chequer boards are used as models for social segregation; a swarm of imaginary billiard balls serves as a model of a gas; and the mathematical structure of a Hilbert space is employed to model a hydrogen atom. None of these objects look anything like the things they are models for, and, more generally, there is nothing in their intrinsic features that would make them models, let alone models for something in particular. Most hydraulic machines are just a piece of plumbing; they aren't models for anything. (Nguyen and Frigg 2022, 55)

We can add the Markov blanket formalism to this list. As I mentioned in the introduction to this book, most scientific models have no concrete existence in nature. They are imaginary constructs, making use of idealizations. The function of "interpretation" is to turn such idealized structures into models of target systems in the world, for example by interpreting some of their features as sensory states, active states, and thus like.

This brings up an important issue to which I now turn: *approximation*. As I shall understand it, the Markov blanket formalism itself is not literally true of target systems. It is an approximate way of representing the boundaries of target systems (following on from chapters 6–7). The scientific realist may endorse the claim that the theoretical description is literally true, but only approximately true of the target system (following on from chapters 3–4). This leaves it open to provide further details about the model and its relation to target systems. I now consider this issue with respect to Bayesian inference and generative models in FEP models.

9.5 Approximate Bayesian Inference

It is common to read that instrumentalism about FEP-models is motivated by appeal to free energy minimization being an approximation of Bayesian inference (see, e.g., van Es 2021).

The usual proposal is that free energy minimization can be shown to be equivalent (under certain mathematical assumptions) to Bayesian inference given a generative model. A generative model is a probabilistic specification of how some kind of data might have been generated. According to Parr and Friston, a generative model "expresses prior beliefs about unobserved hidden states [i.e., causes of sensory input], the probabilistic dependencies between these states, and a likelihood function that maps hidden states . . . to sensory data" (2018, 2). A generative model can be used to predict new data. It can be used to infer the hidden states that may have caused the observed data (Beal 2003). This allows a Bayesian rendition of active inference under the FEP; namely, that internal, sensory, and active states of an organism can be understood as engaging in optimizing posterior beliefs over a (generative) model as new evidence (sensory input) is generated.

Here is a very familiar point by now. All computational models of the mind/brain are simplified, idealized and inexact models, including the FEP. In systems neuroscience, most models are based on the efficient coding hypothesis. A version of this is the FEP. The efficient coding hypothesis (discussed in chapter 8) proposes that neurons are optimized to maximize information about the internal and external environment. One can model this by introducing a mathematical object known as a generative model, as is done in the FEP. This is the strategy used by Bayesian computational models. Generative modeling is an unsupervised learning method in machine learning. It enables systems to discover and learn about regularities or patterns in input data. Hence, a generative model can be used to give a probabilistic description of how some type of data (e.g., sensory input) might have been caused. Generative modeling is now heavily used in neuroscience, especially because the field is moving to an understanding of the brain as a probabilistic organ. The is precisely what is captured by the FEP. Generative models are idealizations. A generative model discovers the joint probability between data and causes by using Bayes's rule (or other approximation methods). Because of this, generative models cannot be literally true of the brain or the mind (Kirchhoff, Kiverstein, and Robertson, forthcoming). Like point objects and infinite populations, Bayes' rule is not literally encoded in the brain. Hence, the brain is not literally a generative model.

To the extent nervous systems are engaged in any kind of Bayesian inference, such systems approximate Bayesian inference—that is, the brain does not literally do inferences. To suggest otherwise would be a clear case of conflating the mathematics of the model with the target system (see, e.g., Andrews 2021). Hence, the FEP, even when given a Bayesian articulation, is not the claim that organisms perform optimal or literal Bayesian inference to minimize surprise. The FEP is the claim that the minimization of free energy conforms to an approximation of Bayesian inference. This is very much how Hohwy approaches the issue here:

> The key point is that approximate inference is not just like exact inference except with approximate values, rather the "approximation" is the minimization of the KL-divergence [the difference between posterior and recognition densities that is minimized during the minimization of variational free energy]: an approximation of the states of the system to the states that it would have if it were indeed computing exact inference. This means that it is correct to say that such a system is literally doing "approximate inference" even if it does not literally do "inference" in the sense of "computations over probability distributions." (Hohwy 2020, 17)

Hohwy's point is that it can be true to claim that brains approximate Bayesian inference even if brains do not literally compute belief updating schemes such as Bayes's rule. Hence, Hohwy's (2020) position is consistent with the conclusions I drew about computational theories of cognition in chapters 6–7. This is also the relationship one finds between natural gasses and ideal gasses, where using the ideal gas model to say something about natural gasses amounts to an approximation. The implication with respect to the FEP is that if both neuronal dynamics and the scientific model succeed in reducing surprise, they both converge on the following: the ability to keep a low conditional entropy over their constituent states. This connects with Hohwy's claim that "it is correct to say that such a system is literally doing 'approximate inference' even if it does not literally do 'inference' in the sense of 'computations over probability distributions'" (2020, 17).

9.6 The Representational Status of the FEP

When developed as a process theory, the FEP takes the form of a computational theory of brain structure and function. This is the case when the

FEP is given in the form of a *prediction error minimization* (PEM) framework. When this is done, the brain is modeled as a hierarchical generative model for minimizing a prediction error quantity (surprise), which, as we know, specifies how the divergence between expected input and actual input relative to an internal model. PEM involves several idealizations, such as that belief updates are performed by computing Bayes's rule and state-transitions are modeled as *partially observable Markov decision processes*.

A brief point of clarification on models is in order now, since we are now talking of two different kinds of models. The sense most familiar to readers of this book, and in general, will be a scientific model: an indirect representation of some possible or actual system. By contrast, in the FEP literature a narrower sense of model is also used, which is called a generative model (Mann, Pain, and Kirchhoff, 2021). As I said above, a generative model is a mathematical entity with applications across statistics and other sciences such as theoretical neuroscience. A generative model will usually take the form of a probability distribution, such as $p(s,\vartheta \mid m)$. There is a serious discussion in the field about how to assess the metaphysical status of such generative models. Some posit that generative models are possessed by agents, who in turn, employ them to perform inference and action. This conforms to a literalist reading of such a concept. Others use generative models to describe the dynamics of agents. It should come as no surprise that I treat talk of generative models nonliterally. That is, discourse about generative models is useful but it should not be taken too seriously, because it is not true that brains literally encode the structure and function of a generative model.

When the FEP is formulated as PEM, it usually takes on the first reading. A literalist take on the ontological status of generative models is that such models are encoded in the nervous system (Kiefer 2017). PEM, therefore, gives special importance to internal, information processing with representational properties. On a PEM view, internal representations are cast in terms of top-down probabilistic inferences on probability density distributions. PEM is a computational framework. Here the computational functions realized by the brain are cast in terms of minimizing prediction error, reflecting the probability of sensorimotor input (or data) conditioned on a statistical (generative) model. On this view, the brain is conceived as realizing the functional roles required to encode a distributed hierarchical Bayesian generative model (Hohwy 2012).

Many take the FEP, when understood as the process theory such as PEM, to be a genuine case of a representational-using system. It is an ontological commitment of PEM that nervous systems manipulate representations. However, the underlying mathematics of the FEP militates against this idea. I do not dispute that the FEP can be given a treatment in terms of PEM. However, the underlying mathematics remains the same.

What does the FEP qua its status as an idealized model tell us about representation? The FEP starts from the close to tautological notion that an adaptive agent must do work to avoid terminal phase boundaries. Parr and Friston provide the following simple example: "An example of a phase boundary, which illustrates this point clearly is a cliff edge. For a flightless organism, structural integrity depends on keeping to one side of this boundary" (2018, 4). How can an organism do this? According to both the FEP and PEM, to avoid such terminal phase boundaries, it is necessary that an organism minimizes the dispersion of observable outcomes—for example, is it far away or too close to the cliff edge. Again, for both the FEP and PEM, the information that an organism has access to are carried by its internal and sensory states. Here is the point. Mathematically, the dispersion of observable states, $\tilde{o}$, is expressed as their Shannon entropy (Parr and Friston 2018, 4):

$$H[P(\tilde{o})] = -E_{p(\tilde{o})}\,[ln\,P(\tilde{o})].$$

Shannon entropy is the information theoretic foundation for the FEP qua mathematical model. It is a quantitative measure of uncertainty or surprise. It is generally agreed that the sense of information developed by Shannon and used in mathematical information theory is both useful and legitimate and that it is relevant for many parts of theoretical neuroscience. It basically says that *X* is a source of information if it has a range of possible states or values, and that *X* carries information about *Y* to the extent that *X* and *Y* are (lawfully, or causally, or physically, or probabilistically) correlated. PEM is formulated as making use of a much richer notion of information: representational content. However, the mathematical formulation of the FEP does not rest on representational content. Rather, it is described exclusively in terms of information such as covariation and mutual information. Hence, even if the FEP is an idealized model, its mathematical formalisms do not support the additional claim that surprise minimization is carried out through top-down probabilistic and representational processes on probability density functions.

One could try to push back against this upshot. For example, in their collaborative work, Kiefer and Hohwy (2018) argue that it is possible to identify a measure for misrepresentation in PEM. They start from the familiar observation that the brain approximates Bayesian inference. Similarly, from Bayes's rule we know that $p(h|e)=p(e|h)p(h)/p(e)$. We also know from machine learning that this problem can be calculated or tackled directly, but is better handled by a system adopting an approximate probability density distribution, $q(h)$, and then trying to make it as close as possible to the true posterior density distribution, $p(h|e)$. On the assumption that the brain is an approximate (variational) Bayesian inference machine, it follows that the internal states comprising the brain can approximate the posterior density, and thereby account in a strictly probabilistic fashion for the causes of its sensory observations. Approximating the posterior density is possible, as the posterior density is equal to the conditional probability of the sensory evidence given a hypothesis. This means that by engaging in approximate Bayesian inference, internal brain states can update the posterior under a generative model by making (implicit) use of Bayes's rule. This strategy is called Bayesian belief parameterization. Kiefer and Hohwy claim that the "measure of success for such a strategy is the Kullback-Leibler divergence, or relative entropy, between $q(h)$ and $p(h|e)$" (2018, 18).

The KL divergence is also part of the mathematical package provided by the FEP. See part "b" of figure 9.1, perception optimizes predictions. This is initially a promising proposal for the following reason: a relation of representation is nonsymmetric; or, asymmetric. The KL divergence is non-symmetric in the sense that $KL(p||q) \neq KL(q||p)$. Do appeals to the KL divergence help ground a notion of misrepresentation. Elsewhere, I have argued that it does not (Kirchhoff and Robertson 2018). Here are the key parts to the argument.

The long-term average of prediction error is entropy. In information theory, the entropy of a random variable, X, with a particular distribution, p, denoted by $H(p)$, is a measure of the uncertainty associated with the distribution of that random variable (Murphy 2012, 57). More technically, entropy is a measure of the spread of a distribution. It is only indirectly a measure of uncertainty, in the sense that if a variable has high entropy, it will exhibit high uncertainty. This means that entropy reflects how surprising a set of outcomes or observations are on average. In the context of PEM, minimizing prediction error can thus be understood as the process of optimizing evidence for a model. Hohwy (2016) aptly summarizes this as self-evidencing. Formally, self-evidencing is the requirement for any biological system to minimize the Shannon entropy of the observations, $\tilde{o}$, it makes:

$$H[P(\tilde{o})] = -E_{p(\tilde{o})}\,[ln\,P(\tilde{o})].$$

This equation shows that Shannon entropy, $H[P(\tilde{o})]$, quantifies the uncertainty associated with a dispersion of outcomes or observations, which we represent as $\tilde{o} = [o1, o2, \ldots on]$. This means that Shannon entropy is the same as the negative expectation under the distribution p over ($\tilde{o}$). Intuitively, the higher the average number of observations (or bits) required to characterize the dispersion of states, the higher the entropy of p. Conversely, if the dispersion is low, p has low uncertainty or unpredictability.

We want to know more than just the average Shannon entropy associated with the distribution of a single random variable. We also want to know the *relative (Shannon) entropy* between the distributions of two random variables. This is evident when we are addressing the KL divergence, which is a bivariant relation between two variables and their probability distributions, $q(h|e)$ and $p(h|e)$. Crucially, relative (Shannon) entropy is the distance or divergence of the KL divergence:

$$q(h|e) \text{ and } p(h|e),\ \mathrm{KL}(q(h)||p(h|e)).$$

In statistics, relative entropy $D(p||q)$ reflects the inefficiency of taking for granted q when the actual (or true) distribution is p, and is written accordingly (Cover and Thomas 2006; see also Kiefer and Hohwy 2018, 18):

$$D(p||q) = \sum p(x)\log(p(x)/q(x)) = -\sum p(x)\log(q(x)/p(x)).$$

Relative entropy is therefore a measure of how close or how far apart the distributions of each variable are. However, relative Shannon entropy is not representational content. There are at least two reasons for this. First, relative entropy is a case of mutual information. Mutual information is an information-theoretic measure of how much information one variable reflects about another variable. Mutual information is the KL-divergence between the distributions $q(h)$ and $p(h|e)$. Second, mutual information is a relation of covariance. A covariance relation is a bivariant relation between two variables. In the case of the KL-divergence we can ask: what happens if the Shannon entropy of one variable (or distribution) goes up or down? If both variables (or distributions) follow the same pattern such that when

$q(h)$ has low entropy, $p(h|e)$ has low entropy, then both distributions covary. Finally, covariance is not any kind of representational content as I argued in chapter 5. Therefore, the mathematical basis of the FEP gives no justification for the RTM.

9.7 Conclusion

This chapter has integrated several core themes covered in this book by addressing a particular case study: the free energy principle (FEP) in theoretical neuroscience. I highlighted the unifying ambitions of the FEP. I have shown in which senses the FEP is an idealized and abstract model. It is a mathematical model that is not literally true of its target domain. However, as with any other idealized model in model-based science, this is no justified reason do deny scientific realism given such models. I considered whether the FEP lends support for the RTM. I did this by looking specifically at the mathematical formalism of the FEP and concluded that one finds no support for the RTM when looking at the mathematical basis of the FEP. Finally, because brains do not literally compute the variational functions of the FEP, it does not lend support to the CTM. The FEP provides us with a computational framework that can be used to make testable inferences about target systems.

10 Idealization and Fictionalism in Science and Cognitive Science

Overview: The nonliteral view of idealization that I have defended in this book does not equate to fictionalism. You might have thought otherwise, since fictionalism about any given domain is typically expressed as the view that discourses within a certain domain are not directed at being literally true. Throughout this book, I have defended, in various ways, a nonliteral view of idealization and I have put it to use to defend a version of scientific realism. I have used this nonliteral view of idealization to examine a number of foundational topics in cognitive science, and drawn some relevant implications on this basis for discussions about neural representation, neural computation, and the prospects of explanatory unification. I have defended the nonliteral view of idealization as implying that idealizations are neither true nor false with respect to target systems, since the function of idealizations is not to represent aspects of target systems in the world. So, how could the nonliteral view of idealization not be a species of fictionalism? That is the topic of this final chapter.

10.1 Introduction

It is easy to understand why one might interpret idealizations as useful fictions. Prima facie, therefore, it makes sense to equate idealization with fictionalism. Consider how Toon introduces fictionalism in his book *Mind as Metaphor*:

> Fictionalism is now a popular position in many fields, including metaphysics, moral philosophy, philosophy of language, and philosophy of mathematics. The core idea of fictionalist approaches is that talk in some area can be useful even if it is not straightforwardly true. For example, a realist about mathematics might argue that mathematical claims like "2+2=4" are true if they correctly describe a

> timeless, Platonic realm of mathematical entities, such as numbers or sets. By contrast, a fictionalist about mathematics might argue that this Platonic realm does not exist and therefore that, strictly speaking, mathematical claims are false. And yet, the fictionalist insists, mathematical claims can still be useful. (Toon 2023, 5)

Godfrey-Smith (2009) asks what the nature of a scientific model is. He defends the idea that scientific models are akin to fictional objects introduced in novels and other forms of literature. Godfrey-Smith states:

> Then we find that, at least much of the time, model systems are treated as fictional systems that lack actual existence but are, in a sense, of the same kind as the target systems that the models are used to help us understand. Often the concern is with sets of cases rather than single cases, on both sides, but each model system itself is something that would be concrete if real; it would be an arrangement of physical entities, located in space and entering into causal relations. I don't claim it is always that way. There is a variety of styles within model-based science, and this includes a sort of "gradient of abstraction" with respect to the handling of models. Some of the variety will be discussed later in this section, but a handling of model systems as imagined concrete things, as fictional concreta, is a clear feature of much of this work. Model-based science often presents itself as an investigation of imagined simpler analogues of a target or targets. This suggests an analogy between model systems and fictional objects introduced in novels and other forms of literature. (Godfrey-Smith 2009, 104)

Both Godfrey-Smith (2009) and Toon (2023) reference Vaihinger's (1924) *The Philosophy of "As If"*. Following Toon (2023), we can start by noting that Vaihinger introduces a distinction between *fictions* and *hypotheses*. On Vaihinger's account, a fiction is false and it is taken to be false by those making use of it. This suggests that the function of a fiction is not to be true of the natural world; rather it is to be useful. A hypothesis, by contrast, makes a particular claim about the world, in the sense that it is a truth-conditional assertion regarding some possible aspects of reality. Hypotheses, on Vaihinger's account, are claims about the world that one might be able to settle by future inquiry. Interestingly, Toon considers the billiard ball model of gases, and states:

> The model makes use of many fictions, in Vaihinger's sense. It assumes that the molecules of the gas collide perfectly elastically, for example, and that they exert no forces on each other between collisions. These assumptions are false and known to be so. And yet they are also enormously useful, since making these assumptions allows us to construct our model of the gas and its behaviour. (Toon 2023, 76)

This clearly expresses the view that fictions are idealizations, and that idealizations conform to fictionalism. Notice, however, how the fictional or idealized posits of the billiard ball model of gases are framed. They are defined as useful falsehoods. Fictions or idealizations are taken by Toon and Vaihinger to be false with respect to aspects of target systems in the world. I called this view the "falsehood view of idealization" (FvI) in chapter 2. In this book, I have defended a nonliteral view of idealization which I have argued does not treat idealizations as falsehoods. Like Toon and Vaihinger, I take idealizations to be useful and legitimate parts of scientific practice. However, unlike them, I reject the view that idealizations are falsehoods. Thus, the first reason why the nonliteral view of idealization is not fictionalism is that I explicitly reject the idea that idealizations are false of target systems. Toon (2023) makes a further important comment on the billiard ball model of gases:

> The model also involves a number of hypotheses. For example, it claims that the gas is made up of molecules and that they exert pressure when they collide with the walls of the container. These are claims about the gas and its behaviour that might have turned out to be true or false—although, in this case, we believe that these hypotheses are true. (Toon 2023, 76)

Basically, what Toon ends up with is the view, which I introduced in chapter 4, that scientific models—in this case the billiard ball model of gases—comprise two types of claims: (a) idealizing, noncorresponding claims; and (b) nonidealizing, corresponding claims. Where I deviate from Toon and Vaihinger is that they both assume that type (a) claims are useful falsehoods. They are, therefore, assuming the FvI. This is important, because type (a) claims come out as falsehood only if taken literally. In chapters 3–4, and across chapters 5–9, I argue that type (a) claims should be taken nonliterally. In chapter 4, I motivated this in the following way. The way I understand type (a) claims can be captured in the form of type (a*). A scientific model is partly composed of a type (a*) claim, where such claims should be treated nonliterally. Type (a*) claims, therefore, do not have a representational function from a model to a target system in reality. This is a second reason for why I do not think that it is justified to equate idealization with fictionalism.

A notable property of fictions is that even if fictions are false, they may nonetheless possess significant (cognitive) value. This is hard to deny.

Consider, for example, the following three cases provided by Kroon, Brock, and McKeown-Green in their *Critical Contribution to Fictionalism*:

Case 1: When I read my children a story about three little pigs who live in houses of straw, sticks and bricks and a big bad wolf who takes pleasure in huffing and puffing to blow their respective houses in, my purpose is to entertain and delight. My aim is not to persuade my children that I believe what I say or that they should believe it either. And my children understand that as well as I do. In such circumstances, then, I am saying something meaningful but I am not asserting what I say, and consequently am not representing myself as believing such things. Instead, I am pretending to assert or, alternatively, telling a story—where this is a distinct kind of speech act. (Kroon, Brock, and McKeown-Green 2019, 3)

Case 2: When I set my students a quiz on Shakespeare's Macbeth, I might ask them a question such as: "Macbeth was told he would be king by three witches he met on the battlefield—true or false?" Any student of English literature should know that the correct answer to this question is TRUE. The play opens with the witches agreeing to meet again—upon the heath to meet with Macbeth. In the third scene they meet with Macbeth and Banquo, and the third witch proclaims, "All hail, Macbeth, that shalt be king hereafter!" It would not be appropriate in an English literature class to complain about being marked down for answering FALSE to the question on the grounds that there are no witches, so Macbeth could never have spoken with one. In this context, the question is clearly about the content of the play, not about the real world. When we engage with fiction in this way, we shouldn't take what's said at face value. Instead we should interpret the questions and claims made in some other way, for example as including a silent story operator of the form "according to such-and-such story." (Kroon, Brock, and McKeown-Green 2019, 3)

Case 3: Emily Bronte's poem "Love and Friendship":

Love is like the wild rose-briar,
Friendship like the holly-tree—
The holly is dark when the rose-briar blooms
But which will bloom most constantly?
The wild-rose briar is sweet in the spring,

Its summer blossoms scent the air;
Yet wait till winter comes again
And who will call the wild-briar fair?

> Then scorn the silly rose-wreath now
> And deck thee with the holly's sheen,
> That when December blights thy brow
> He may still leave thy garland green.
> (Kroon, Brock, and McKeown-Green 2019, 3–4)

Case 3 is a clear-cut example of how figurative language possesses significant cognitive value. As Sprevak puts it: "One value of figurative language is that it provides an alternative way of expressing claims than literal assertion" (2013, 4). Case 3 makes use of similes such as comparing love to the wild rose-briar and friendship to the holly tree. Similes are figurate forms of language that should not be taken literally: love is not literally a wild rose-briar and friendship is not literally a holly-tree. Figurative language is, therefore, nonliteral. Fictionalism of this kind shares nonliteralism with the view of idealization that I have been developing in this book. Yet, figurative claims are usually taken to be false even if meaningful. I have defended idealizations as neither true nor false, yet as still being useful and legitimate parts of scientific practice.

I will have the same to say about case 1. So let me say a little more about case 2, since it will allow me to further elaborate on the nonliteral view of idealization. I take it to be obvious that the answer to the question, "Macbeth was told he would be king by three witches he met on the battlefield—true or false?" is TRUE, even if it would be false with respect to actual-world target systems. This turns out to have an important implication for the nonliteral view of idealization that I have been developing in this book. For example, I take the discussion about structural representation to be true in the same sense as I take it to be true that "Macbeth was told he would be king by three witches he met on the battlefield." The same is the case for Shea's (2020, 113) model of structural correspondence. We can ask: does I map V to H in virtue of structural relations between $v_1 \ldots v_4$ to $x_1 \ldots x_4$? The answer to this question is TRUE. However, the mapping relation described by I is an idealization (from chapter 5). If taken literally, it is false across the model-world relation. If taken nonliterally, then one option is to say that I is neither true nor false—again when the comparison is made from a model to reality. I also consider the normalization law discussed in chapter 7 as true:

$$R = \gamma \frac{I^n}{\sigma^n + I_m^n + I^n}.$$

It is true of the scientific model, even if it is an idealization. So again: if taken literally, the normalization equation is false when considered across the model-world relation; however, if taken nonliterally, the normalization equation need be neither true nor false when considered across the model-world relation. The reason is, as we know, that the function of idealization is not to represent aspects of target systems. This lends additional support to the claim that the nonliteral view of idealization is not equal to fictionalism.

In the rest of this chapter, my aim is to elaborate on the claim that my defense of a nonliteral view of idealization is no defense of fictionalism. I will do this by addressing some of the main topics covered in the book. In section 10.2, I look at idealized models and fictionalism, leaning on Godfrey-Smith (2009) and Frigg (2010). In section 10.3, I consider recent proposals about mental fictionalism and fictionalism about neural representation, drawing on the work of Toon (2023) and Sprevak (2013). In section 10.4, I will make a few remarks about the computational theory of cognition, idealization, and fictionalism, building on material introduced in the book on these topics. I conclude in section 10.5 with a discussion on the relationship between scientific realism, fictionalism and idealization, drawing on Suárez (2010).

10.2 Idealized Scientific Models and Fictionalism

According to Frigg, scientific models are "akin to places and characters of literary fiction, and that therefore theories of fiction play an essential role in explaining the nature of model-systems" (2010, 99). Godfrey-Smith (2009), as we saw above, endorses the same kind of view. He tells us that model-based science often presents itself "as an investigation of imagined simpler analogues of a target or targets" (2009, 104), thereby supporting the analogy between scientific models and fictional objects. Middle Earth does not actually exist. However, Middle Earth would be concrete if real. The world of a novel, Godfrey-Smith says, is a candidate for physical existence: "If real, it would be an organized system characterized by causal relations among its components. Such fictional objects, both in literature and science, can often be mathematically described, but that does not imply that the fiction is itself a mathematical object" (2009, 104). It is precisely because one can describe scientific models in concrete and mathematical terms that they can be used effectively to reason about the world, even if such models have the status of Middle Earth in *The Lord of the Rings*. Here is what I take to be a representative statement about fictionalism concerning (idealized) scientific models. According to Frigg:

> The core of the fiction view of model-systems is the claim that model-systems are akin to places and characters in literary fiction. When modeling the solar system as consisting of ten perfectly spherical spinning tops physicists describe (and take themselves to be describing) an imaginary physical system; when considering an ecosystem with only one species biologists describe an imaginary population; and when investigating an economy without money and transaction costs economists describe an imaginary economy. These imaginary scenarios are tellingly like the places and characters in works of fiction like Madame Bovary and Sherlock Holmes. These are scenarios we can talk about and make claims about, yet they don't exist. (Frigg 2010, 101)

I do not think that idealizations in science are like places and characters in literary fiction. Galileo's frictionless plane was an imaginary physical structure, and one that Galileo sought to realize, at least approximately. Idealizations are clearly imaginary, and they require an extensive skill-set to use efficiently. In other words, I do not think that the normalization law or the variational free energy notation in the FEP or the HH model or the Schelling model or the Hardy-Weinberg model are akin to the places and characters in *The Lord of the Rings*. As I argued at the end of chapter 3, idealizations can be given a precise mathematical formulation. They are real parts of scientific models—abstract tools used to explain reality. Moreover, idealizations such as normalization law and the variational free energy notation are the results of years of painstaking experimental work in neuroscience. They are both meaningful and useful. They are imaginary constructs with a mathematical reality. However, unlike fictions, I have defended idealizations in science and cognitive science as neither true nor false about the world. As I concluded above, this is one reason why I reject fictionalism about scientific models.

10.3 Fictionalism about Representation

Toon (2023) defends what he calls *mental fictionalism*. His primary target is folk psychology about the mind. Mental fictionalism is motivated by the observation that talk about mental properties—aboutness, semantics, misrepresentation, and so on—applies first and foremost to public representations;

in particular, these properties apply to language. Mental fictionalism, as defended by Toon, is the view that when we engage in discourse about mental states we "transfer these semantic categories metaphorically to the mind as an inner realm" (2023, 63). That is, we talk as if people have such semantic categories inside their heads, without taking discourse about mental states literally. In this book, I have had nothing to say about the folk psychological view of the mind. Rather, my focus has been explicitly on neural representation because work on neural representation is, to a large part, based on modeling work in computational neuroscience—the latter being a nice example of model-based science. I will therefore sidestep discussion of this view of representational fictionalism (for other treatments of fictionalism about folk psychology, see Demeter 2013; Wallace 2016). The common thread between Demeter (2013), Toon (2023) and Wallace (2016) is that folk psychology is false, but useful for "prediction, understanding, and normative evaluation" (Sprevak 2013, 3).

Before moving onto discussing neural representation, it is well worth our time to stop, even if shortly, and consider Dennett's (1987) "three stances." The first is the *physical stance*, where one might seek to explain some observed behavior by reference to physical laws. One might adopt this stance to explain and predict an object's acceleration conditioned on having knowledge of its mass and so on. The second is the *design stance*, which is the natural stance to adopt when we are confronted with manufactured things. The last is the *intentional stance*, which is the sort of strategy that we adopt when trying to make sense of other people's behavior. By adopting the intentional stance, we are treating others as if they have beliefs and desires, and so on, for the purposes of explaining and predicting their behavior. According to Dennett's position, one should treat folk psychological posits as *abstracta*, i.e., as concepts such as centers of gravity. This is the conclusion I made about neural representations in chapter 5. Dennett explicitly says of his theoretical stances that they do not imply any kind of fictionalism about some domain of discourse, such as folk psychology. A specific reason for this is that Dennett takes fictionalism to endorse the view that theoretical claims about some domain of discourse is false even if useful (1987, 72). This is also how fictionalism is defined by Toon (2023) and Vaihinger (1924), both of whom take fictions to be useful falsehoods. I argued above that the nonliteral view of idealization is not a species of fictionalism because it rejects the view that idealizations are useful falsehoods. Here is how the nonliteral view about idealization differs from Dennett's position. In making his position explicit, Dennett relies on a parallel with centers of gravity:

> I should have forsworn the term [instrumentalism] and just said something like this: My *ism* is whatever *ism* serious realists adopt with regard to centres of gravity and the like, since I think beliefs . . . are *like that*—in being *abstracta* rather than part of the 'furniture of the physical world' and in being attributed in statements that are *true* only if we exempt them from a certain familiar standard of literality. (Dennett 1987, 72)

Dennett, therefore, endorses the view that folk psychological posits, and idealized posits such as centers of gravity, can be true even if none of these posits literally exists. Centers of gravity are imaginary points. Formally, the location in space-time where weight is equally distributed in all directions is what is known as the center of gravity. Centers of gravity are, therefore, idealizations—in the same sense that infinite populations are idealizations. Dennett says that these idealized posits are true and that folk psychological posits are true. Yet he adds an important qualification. According to Dennett: "Each is a truth one must understand *with a grain of salt*" (1987, 72–73). I do not consider idealizations as being true in the sense of being truths with a grain of salt. If we view idealizations from within a model, they can often be true (and not just with a grain of salt). For example, the mathematical notations underlying Jensen's inequality—which can be used to show that variational free energy is an upper bound on entropy is true. In Kirchhoff et al. (2018), we provide the proof (created by Thomas Parr) shown in figure 10.1.

The explanation for this is as follows. First, the curve of a logarithm function is concave, which has implications when averages are taken either of the logarithm or of the input to the function. Second, figure 10.1 considers a dataset of two points, x_1 and x_2. The x-axis shows that the average of these points will lie halfway between x_1 and x_2. Third, when the logarithms of x_1 and x_2 and their average are calculated, the log of the averages lies above the midpoint between $\log x_1$ and $\log x_2$. This is shown on the y-axis. Since the midpoint is the average of the logarithms, this implies that the log of an average is greater than the average of a log. This is Jensen's inequality. Finally, if one replaces the variable x with a ratio of probabilities,

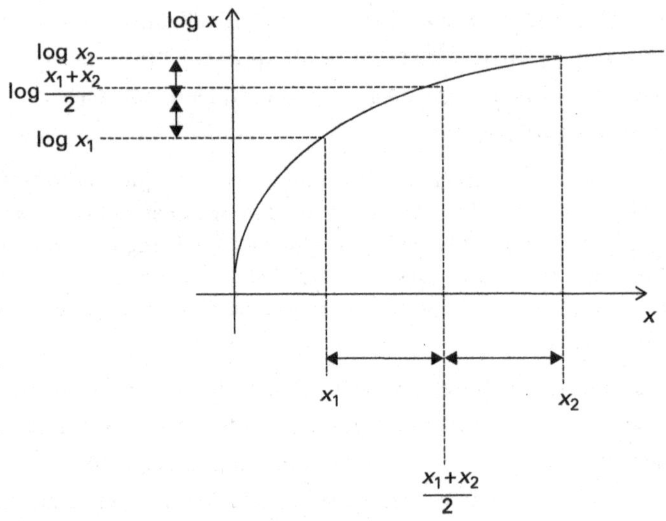

Figure 10.1
Illustration of Jensen's inequality, showing that free energy is an upper bound on surprise.

$P(o,s)$ / $Q(ss)$, and take the averages with respect to a distribution $Q(s)$, then one can use Jensen's inequality to write:

$$E_{Q(s)}\left[ln \frac{P(o,s)}{Q(s)}\right] \le ln\ E_{Q(s)}\left[\frac{P(o,s)}{Q(s)}\right].$$

The term on the left represents the negative free energy, while the term on the right represents the negative surprise. To illustrate this, one can write this out in full as follows:

$$E_{q(s)}\left[\frac{P(o,s)}{Q(s)}\right] = \Sigma_s \frac{Q(s)P(o,s)}{Q(s)} = \Sigma_s\ p(o,s) = P(o).$$

This allows one to use Jensen's inequality to show that the free energy is an upper bound on surprise as follows:

$$F \ge -ln\ P(o).$$

This is a true statement of the claim that free energy bounds surprise, and it can be shown to be formally accurate by Jensen's inequality. Idealizations can therefore be true within a model. The nonliteral view of idealization is the view that idealized posits are neither true nor false when compared across the model-world relation.

In the rest of this section, I focus on Sprevak's (2013) account of *neural representation fictionalism* (NRF). Appeals to neural representations are everywhere in computational neuroscience, and are ascribed to the nervous system at all levels of processing, from low-level sensory and motor processing to high-level planning and inference. In chapter 5, I motivated the idea of neural representation with the discovery of place cells in the rat hippocampus and their (alleged) role in spatial representation and navigational behavior. As I argued in chapter 5, realism about the RTM based on neural representations is based on the commitment to the actual existence of neural representations. Sprevak asks: "How do neural representations fit into the physical world? What elevates certain neural states to be representations?" (2013, 8). He quickly rules out two answers: (1) it will not work to point out that neural states get to be representational states by social convention; and (2) mere appeal to the response-selectivity of a neural state to a given stimulus will also not work, because response-selectivity is insufficient for something to be a neural representation. As Sprevak notes: "Neural representations are claimed to achieve this feat by being natural representations" (2013, 8).

Developing an account of natural representation requires naturalizing explanations of neural representation. However, all attempts at naturalizing representation have failed, for different reasons.

Two immediate options can be found in the literature. One is "hard-headed realism" (Sprevak 2013, 9). This is the view that we continue to talk about neural representation as being fact-stating and mind/observer-independent properties about nervous systems. The situation is akin to discussions about physicalism. Physicalism must be true, even if we do not have conclusive evidence for this thesis yet. But, future research will prove it to be true. The other option is "eliminativism" (Sprevak 2013, 9). This is the view that we should eliminate from scientific discourse appeals to neural representations, since these theoretical constructs are not fact-asserting about mind/observer-independent properties of nervous systems. Both options are controversial. Hard-headed realism must tackle the task of naturalizing representation. A task that many deem not plausible. Eliminativism necessitates a complete overhaul of cognitive science, as we know it. Fictionalism offers a third option. Sprevak defines NRF as follows:

> According to NRF, neural representation talk is false but serves an important purpose and, for that reason, should be preserved. NRF claims that

> neural-representation talk in cognitive science is perfectly in order and cannot, and should not, be eliminated or paraphrased away from serious fact-stating language. However, neural-representation talk does not bring with it any commitment to the existence of neural representations since it is understood as systematically false. (Sprevak 2013, 10)

Here are the central differences between NRF and what I said about neural representation in chapter 5. I argued that neural representations are best understood as kinds of idealization in the modeling of the mind. In chapter 5, I applied the nonliteral view of idealization to an evaluation of the status and standing of neural representation. I argued that viewing neural representation through the lens of the nonliteral view yields three implications for neural representations qua idealizations: (1) discourse about neural representations is neither true nor false, epistemically useful, and cannot be eliminated; (2) discourse about neural representations is neither true nor false, epistemically useful, and can be eliminated; and (3) discourse about neural representations is neither true nor false, epistemically useful, but are irrelevant for explanation. (1)–(3) all deny NRF, since the NRF is premised on the thesis that all scientific discourse is truth-apt—I called this unrestricted semantic realism in chapter 4.

10.4 Computationalism, Idealization, and Fictionalism

In this section, I will convert the NRF into a view of *neural computational fictionalism* (NCF), mirroring NCF in terms of how Sprevak (2013) formulates NRF. According to NCF, neural computation talk is false but serves an important purpose and should, therefore, be preserved rather than eliminated. Unlike its realist alternatives, NCF does not amount to the view that nervous systems actually compute, since the CTC is understood to be systematically false.

Most of what I have argued in chapters 6–9 does not map onto NCF. Let us start by considering what NCF takes to be false:

1. Edge detection in early vision is realized by computing the Laplacian of a Gaussian distribution.
2. Computation involves a mapping over input, possible internal states, and output in accordance with a rule (such as the normalization law).
3. Entropy is minimized by reducing a variational free energy bound on entropy.

From the perspective of NCF, statements of the form 1–3 are useful and fact-stating. They are useful ways of talking about processes involved in 1–3 even if 1–3 are false. Here is an interesting consequence, also discussed by Sprevak (2013) but for NRF. In the neural computational fiction, the following claims are true:

4. In the neural computational fiction, edge detection in early vision is realized by computing the Laplacian of a Gaussian distribution.
5. In the neural computational fiction, computation involves a mapping over input, possible internal states, and output in accordance with a rule (such as the normalization law).
6. In the neural computational fiction, entropy is minimized by reducing a variational free energy bound on entropy.

Sprevak makes the following claim about NRF: "In familiar fiction, e.g. the Sherlock Holmes stories, there is a written body of text that supplies the fiction. This text determines what is, and isn't, true according to the fiction. For NRF, there is no such text. Instead, the Neural Representation Fiction should be understood as the best agreed theory about what neural states represent. This theory is implicit in the practice of cognitive science" (2013, 11). We can frame the NCF similarly: for NCF, there is no text of fiction which determines what is true and false according to the fiction. Like NRF, NCF should be understood as an agreed theory about computational processes in the nervous system. The theory is implicit in the practice of cognitive science. This is fictionalism about neural computation.

Regarding 1–3, I have argued that 1–3 are all clear-cut candidates of idealizations in science. 1–3 are false *only if* they are understood to be representations of aspects of target systems. However, as I have argued, 1–3 in virtue of their status as idealizations are not false precisely because the function of idealizations is not to represent aspects of target systems. I have argued that those defending realism about computational theories of cognition fall into the trap of confusing the properties of the model with the properties of the target system under investigation. Regarding 4–6, no doubt one can find cognitive scientists adhering to something like the NCF. I do not want to deny this. However, as I have been discussing 4–6, they are true within their own respective scientific model. It is literally true that entropy is minimized by reducing a variational (free energy) bound on entropy. It is true that this is how an FEP model defines entropy minimization. This is not any kind of fictionalism.

10.5 Scientific Realism, Idealization, and Fictionalism

Godfrey-Smith (2009), as we know, thinks that the nature of scientific models is akin to fictions in literature. Yet, as he says: "I also assume, without argument, a kind of scientific realism. By means of theorizing, including the theoretical investigations of fictions of this kind, we learn about the world, and not only about its observable structure" (2009, 102). One reason for why Godfrey-Smith (2009) might think that fictionalism about scientific models is consistent with scientific realism is that in the philosophy of science terms like "fiction" and "fictionalism" are usually taken to be a view about the modeling of model building in science (Suárez 2010).

Fictionalism is often seen as antithetical to realism. For instance, mathematical realism is the view that mathematical truths are objective and mind-independent. As the mathematical realist sees it, mathematics is the study of necessary and unchanging facts, which mathematicians aim to discover. Mathematical fictionalism, by contrast, is the view that there is no reason to suppose that mathematical objects exist in an objective, mind-independent way. Rather, discourse about mathematics is fictional discourse. Fictional discourse is perfectly meaningful and legitimate ways of talking without the need to posit any kind of particular ontology about mathematics.

At this stage it will be instructive to draw a distinction between "fictionalism," on the one hand, and "scientific fictionalism," on the other. In section 10.4, we looked at both NRF and NCF, which both fall under "fictionalism." For example, NCF states that it is false that nervous systems are literally computational systems, whereas it is true that nervous systems are computational systems in the NCF. Presumably this is what allows fictionalists to say that certain posits are false of target systems even if discourse about such posits can be meaningful and useful. As I have argued, I reject fictionalism of this variety, since the nonliteral view of idealization is that idealizations lack truth-values across the model-world relation.

The term "scientific fictionalism" is from Suárez (2010), who states that scientific fictionalism is the view that the creative imagination is indispensable for building fertile models of target systems in the natural, cognitive and social sciences. Note that it does not follow from this that scientific models are imaginary simpliciter. Scientific fictionalism, as I understand it, is the view that creative imagination is an essential feature of building models in the sciences. Unlike "fictionalism," therefore, scientific fictionalism is not motivated by appeals to metaphysical parsimony. It is about the methodology of model building.

Let us make the distinction between fictionalism and scientific fictionalism a little sharper by comparing these terms to "realism." In doing so, I draw on Kroon (2011). Kroon invites us to think of realism about a certain area of discourse *D* (e.g., scientific discourse). Realism about *D* claims (1) that sentences of *D* have truth-conditions, in the sense that they have a representational semantics, and (2) that aiming at (approximate) truth are assertive utterances and epistemic acceptance when engaging in *D*. Veritism is, therefore, the norm for epistemic acceptance of discourse in *D*. According to Kroon, fictionalism about sentences of *D* accepts (1) but denies (2). For defenders of fictionalism, the "norm of acceptance is not truth—one can accept a sentence of *D* for the best of reasons, justifiably acting on it, using it in one's theorising, drawing inferences and acting on those, and so on, but not actually believe it, since there are benefits other than truth for whose sake one should accept such a sentence" (2011, 787). Kroon provides the following example of fictionalism, from chapter 1 of Jane Austen's *Pride and Prejudice*: "Mr Bennet was so odd a mixture of quick parts, sarcastic humour, reserve, and caprice, that the experience of three and twenty years had been insufficient to make his wife understand his character" (quoted in Kroon 2011, 788).

The example supports (1) about *D*, since Austen aims to describe the world as being a certain way—as ascribing a set of properties to an individual named "Mr Bennet." However, as Kroon notes, Austen is not aiming to report that a person who actually exists has this set of properties. Rather, she is pretending that a person exists who has this set of complex properties. As Kroon puts it: "In fiction we find the combination of a representational semantics and a notion of acceptance that is not truth-normed" (2011, 788).

Fictionalism, I take it, is hostile to realism about *D*. Or, fictionalism is hostile to realism about *D* on the standard view of realism, where I shall treat "realism" as "scientific realism." It is an open question whether fictionalism is unfriendly to all forms of scientific realism. Potochnik, for instance, endorses the view that realism "without a strict commitment to truth better accommodates the nature of our present-day scientific achievements, and it opens up an interpretation of cumulative scientific progress that resists the antirealist's pessimism" (2023, 152). One of the reasons Potochnik provides us in support of this view of realism is that "model-based explanations proceed with substantial assistance from idealizations or even fictions" (153). Moving away from truth as the quintessential aim of science might seem natural if "false posits or fictions facilitate at least some varieties of scientific explanations, this is role is another form of sacrificing some truth to facilitate other aims" (153). On this basis, Potochnik concludes: "Insofar as model-based science is an important form of science, it seems that truth, approximate truth, or increasing truth may not always best describe the accomplishments—or even the aim—of science" (153).

Fictionalism is hostile to scientific realism if scientific realism is defined as having a veritist epistemology. In this book, I have defended veritism as the epistemic aim of science (see chapters 3–4). Therefore, my defense of scientific realism in the context of partial and idealized scientific models is not an example of fictionalism. Nor has my defense of scientific realism any close ties to Potochnik's (2023) realism. In my view, removing truth as the central aim of scientific investigations is to deny realism about science. Furthermore, Potochnik's realism is based on, and motivated by, the FvI: that idealizations are useful falsehoods. Idealization, as I have defined it, can accommodate veritism as the epistemic aim of science insofar as one treats idealization in scientific models nonliterally. A nonliteral treatment of idealization in scientific models expresses a commitment to a modified version of (1) and a full endorsement of (2) about *D*. Hence, the nonliteral view of idealization can (a) be used in a defense of scientific realism and (b) does not conform to fictionalism, as it is defined here.

Scientific fictionalism, however, does not amount to fictionalism and does not make scientific realism controversial. In fact, the nonliteral view of idealization can be understood as a type of scientific fictionalism, but only if scientific fictionalism does not amount to the view that idealizations are falsehoods. That is, there is nothing in the definition of scientific fictionalism that mitigates against the nonliteral view of idealization as I have defended it in this book. In the natural and cognitive sciences, where models are used to explain target systems, there is an undeniable role for the creative imagination. Frontloading the importance of creative imagination is not the same as committing to any kind of unconstrained imaginings (Suárez 2010). For example, in arriving at the view that a variational free energy bounds entropy, scientists working on the FEP are constrained by various relations between variational notations of free energy and thermodynamic free energy, in addition to having to specify how the formal variational notation relates to notions such as Gibbs free energy, and so on. In other words, scientific fictionalism is the view that model building needs creativity to produce fertile models of aspects of target systems. However, it is ultimately a view that insists that creative imagination in model building is heavily constrained.

A final word about "scientific fictionalism" is in order. According to Suárez (2010), an important prima facie consequence of scientific fictionalism is that it does not entail that fecund scientific models must be accurate or true. The reason for this qualification is that idealization is so commonplace with scientific modeling practices that misrepresentation appears to be a precondition for scientific modeling to achieve its aims (Suárez 2010). The nonliteral view accepts that insofar as scientific fictionalism is an indispensable part of scientific model-building, there is no entailment relation between scientific fictionalism and the further requirement that fertile scientific models and/or representations must be true or accurate. This is part of how I defined the nonliteral view of idealization. However, from chapter 3 onward, I have argued that the widespread view of idealization as a form of misrepresentation is false.

Once this move is made, the doors to scientific realism (and not fictionalism) are wide open for modeling the mind and the rest of the natural world. This is what I have aimed to establish in this book.

Concluding Remarks: The Idealized Mind

I have argued that when we seek to explain the nature and function of the mind and brain we inevitably make use of simplified and idealized models to do so. In this specific sense, when attempting to explain the mind and brain, we idealize it in much the same way as we make use of idealizations in other sciences to explain aspects of reality. Cognitive science, like any other model-based science, is an idealized science. This is the picture of our models of the mind brought out in this book.

This book does not present my whole view of the idealized mind. It represents the first of several projects on idealization in cognitive science. I have discussed the role of idealization in models that posit neural representations and neural computations. I have argued for a way of defending the prospects of explanatory unification, even when making use of multiple different and idealized models of the mind. Yet, I have left out other core topics of discussion, primarily due to lack of space. For example, I did not discuss the status and standing of information-processing in cognitive science. I should have because talk of brains as receivers of messages, as conceivers, as encoders and decoders, as message-passing devices, and so on, is theoretically misleading - unless we admit that we are only speaking metaphorically (Sganzerla, Hutto, and Kirchhoff forthcoming). I did not tackle a large number of issues in computational modeling and computational neuroscience. I did not take issue with the many and varied philosophical thought-experiments such as philosophical zombies, twin-earth scenarios, the Chinese Room, Swampman, and so on. Nor did I touch on the many versions of functionalism. I also neglected models of consciousness such as the global workspace theory or integrated information theory - both of which make use of idealizations. The list goes on.

What I hope to have shown is that some of our most foundational concepts in cognitive science are akin to idealized concepts in other sciences such as ideal gases, point objects, infinite populations, centers of gravity, and frictionless planes. What this implies for science, including cognitive science, is that the models we use to study and understand the natural world are not strictly speaking part of the real patterns of reality. Rather, they are abstract explanatory devices that exist (partly) in the imagination of the scientific communities that use them to describe, explain, and predict aspects of the natural world.

I arrived at this picture of model-based cognitive science by first relying on discussions about scientific representation, scientific models and idealization, which I used to develop a view of idealization called the nonliteral view of idealization. I then identified several ways in which concepts such as neural representation and neural computation share the same kind of function as idealizations in other sciences such as physics and biology. Then I used the nonliteral view of idealization to show how it is possible to still defend a realist perspective on scientific theories and models in the cognitive sciences, without having to accept either entity realism, eliminativism or fictionalism about discourse involving appeals to neural representation and neural computation. Finally, I showed how it is possible to combine all these elements in plausible defense of explanatory unification in the sciences of the mind.

References

Abraham, Tara. 2018. "Cybernetics." In *The Routledge Handbook of the Computational Mind*, edited by Mark Sprevak and Matteo Colombo, 52–64. London: Routledge.

Andrews, Mel. 2021. "The Math Is Not the Territory: Navigating the Free Energy Principle." *Biology & Philosophy* 36:1–19.

Appiah, Kwame Anthony. 2017. *As If: Idealization and Ideals*. Cambridge, MA: Harvard University Press.

Batterman, Robert, and Colin Rice. 2014. "Minimal Model Explanations." *Philosophy of Science* 81 (2): 349–376.

Beal, Matthew. 2003. "Variational Algorithms for Approximate Bayesian Inference." PhD diss., University College London.

Bechtel, William, and Robert Richardson. 2010. *Discovering Complexity: Decomposition and Localization as Strategies in Scientific Research*. Cambridge, MA: MIT Press.

Beer, Randall. 2004. "Autopoiesis and Cognition in the Game of Life." *Artificial Life* 10:309–326.

Borges, Jorge Luis. 1999. *Collected Fictions*. Translated by Andrew Hurley. London: Penguin.

Brette, Roman. 2013. "What Is Computational Neuroscience? What Is a Realistic Model?" Last modified November 13, 2013. http://romainbrette.fr/what-is-computational-neuroscience-xx-what-is-a-realistic-model/.

Brochini, Ludmilla, Ariadne de Andrade Costa, Miguel Abadi, Antonio Roque, Jorge Stolfi, and Osame Kinouchi. 2016. "Phase Transitions and Self-Organized Criticality in Networks of Stochastic Spiking Neurons." *Scientific Reports* 6:35831.

Bruineberg, Jelle, Krys Dolega, Joe Dewhurst, and Manuel Baltieri. 2021. "The Emperor's New Markov Blankets." *Behavioral and Brain Sciences* 45:10.1017/S0140525X21002351.

Calcott, Brett. 2013. "Why How and Why Aren't Enough: More Problems with Mayr's Proximate-Ultimate Distinction." *Biology & Philosophy* 28:767–780.

Candan, Gurmukh, and Marco Cascella, M. 2023. "Gas Laws and Clinical Application." National Library of Medicine. Last modified August 28, 2023. https://www.ncbi.nlm.nih.gov/books/NBK546592/.

Cao, Rosa. 2018. "Computational Explanations and Neural Coding." In *The Routledge Handbook of the Computational Mind*, edited by Mark Sprevak and Matteo Colombo, 283–296. London: Routledge.

Cao, Rosa. 2022. "Putting Representations to Use." *Synthese* 200 (2): 1–24.

Carandini, Matteo, and David Heeger. 2012. "Normalization as a Canonical Neural Computation." *Nature Reviews Neuroscience* 13:51–62.

Carillo, Natalia, and Tarja Knuuttila. 2022. "Holistic Idealization: An Artifactual Standpoint." *Studies in History and Philosophy of Science* 91:49–59.

Cartwright, Nancy. 1983. *How the Laws of Physics Lie*. New York: Oxford University Press.

Cartwright, Nancy. 1994. "Fundamentalism vs. the Pathwork of Laws." *Proceedings of the Aristotelian Society* 93 (2): 279–292.

Chakravartty, Anjan. 2001. "The Semantic or Model-Theoretic View of Theories and Scientific Realism." *Synthese* 127:325–345.

Chakravartty, Anjan. 2017. "Scientific Realism." *Stanford Encyclopedia of Philosophy* (Summer 2017 Edition), edited by Edward N. Zalta. https://plato.stanford.edu/archives/sum2017/entries/scientific-realism/.

Chalmers, David. 2011. "A Computational Foundation for the Study of Cognition." *Journal of Cognitive Science* 12:323–357.

Chirimuuta, Mazviita. 2014. "Minimal Models and Canonical Neural Computations: The Distinctness of Computational Explanation in Neuroscience." *Synthese* 191:127–153.

Chirimuuta, Mazviita. 2018. "Vision." In *The Routledge Handbook of the Computational Mind*, edited by Mark Sprevak and Matteo Colombo, 397–409. London: Routledge.

Chirimuuta, Mazviita. 2024. *The Brain Abstracted*. Cambridge, MA: The MIT Press.

Churchland, Paul. 1986. *Neurophilosophy: Toward a Unified Science of the Mind/Brain*. Cambridge, MA: MIT Press.

Churchland, Patricia, and Terrence Sejnowski. 1990. "Neural Representation and Neural Computation." *Philosophical Perspectives* 4:343–382.

Clark, Andy. 2016. *Surfing Uncertainty: Prediction, Action, and the Embodied Mind*. Oxford: Oxford University Press.

Clark, Andy, and Joseba Toribio. 1994. "Doing without Representation?" *Synthese* 101 (3): 401–431.

Coelho Mollo, Dimitri. 2020. "Content Pragmatism Defended." *Topoi* 39:103–113.

Coelho Mollo, Dimitri. 2021. "Deflationary Realism; Representation and Idealisation in Cognitive Science." *Mind & Language* 37:1048–1066.

Colombo, Matteo, and Cory Wright. 2016. "Explanatory Pluralism: An Unrewarding Prediction Error for Free Energy Theorists." *Brain and Cognition* 112:3–12.

Corcoran, Andrew, Giovanni Pezzulo, and Jakob Hohwy. 2021. "From Allostatic Agents to Counterfactual Cognisers: Active Inference, Biological Regulation, and the Origins of Cognition." *Biology & Philosophy* 35:1–45.

Cover, Thomas M., and Joy A. Thomas. 2006. *Elements of Information Theory*. 2nd ed. Hoboken, NJ: John Wiley.

Craver, Carl E. 2007. *Explaining the Brain: Mechanisms and the Mosaic Unity of Neuroscience*. Oxford: Oxford University Press.

Craver, Carl E. 2008. "Physical Law and Mechanistic Explanation in the Hodgkin and Huxley Model of the Action Potential." *Philosophy of Science* 75:1022–1033.

Cummins, Robert. 1996. *Representation, Targets, and Attitudes*. Cambridge, MA: MIT Press.

Da Costa, Lancelot, Thomas Parr, Noor Sajid, Sebastijan Veselic, Victoria Neacsu, and Karl Friston. 2020. "Active Inference on Discrete State Spaces: A Synthesis." *Journal of Mathematical Psychology* 99:102447.

Dale, Rick, Eric Dietrich, and Anthony Chemero. 2009. "Explanatory Pluralism in Cognitive Science." *Cognitive Science* 33 (5): 739–742.

Dayan, Peter, and Laurence Abbott. 2001. *Theoretical Neuroscience: Computational and Mathematical Modeling of Neural Systems*. Cambridge, MA: MIT Press.

Demeter, Tomas. 2013. "Mental Fictionalism: The Very Idea." *The Monist* 96 (4): 483–504.

Dennett, Daniel. 1987. *The Intentional Stance*. Cambridge, MA: MIT Press.

Dennett, Daniel. 1991. "Real Patterns." *Journal of Philosophy* 88:27–51.

Dretske, Fred. 1981. *Knowledge and the Flow of Information*. Oxford: Blackwell.

Dretske, Fred. 1986. "Misrepresentation." In *Belief: Form, Content, and Function*, edited by Radu Bogdan, 17–36. Oxford: Oxford University Press.

Egan, Frances. 2018. "The Nature and Function of Content in Computational Models." In *The Routledge Handbook of the Computational Mind*, edited by Mark Sprevak and Matteo Colombo, 247–258. London: Routledge.

Egan, Frances. 2020. "A Deflationary Account of Mental Representation." In *What Are Mental Representations?*, edited by Joulia Smortchkova, Krzysztof Dołęga, and Tobias Schlicht, 26–53. New York: Oxford University Press.

Eklund, Matti. 2024. "Fictionalism." *Stanford Encyclopedia of Philosophy*, edited by Edward N. Zalta. https://plato.stanford.edu/entries/fictionalism/.

Elgin, Catherine Z. 2010. "Telling Instances." In *Beyond Mimesis and Convention: Representation in Art and Science*, edited by Roman Frigg and Matthew C. Hunter, 1–18. London: Springer.

Elgin, Catherine Z. 2017. *True Enough*. Cambridge, MA: MIT Press.

Elgin, Catherine Z. 2021. "Replies." *Synthese* 198:1577–1597.

Eliasmith, Chris. 2003. "Moving beyond Metaphors: Understanding the Mind for What It Is." *Journal of Philosophy* 100 (10): 493–520.

Elliott-Graves, Andrew, and Michael Weisberg. 2014. "Idealization." *Philosophy Compass* 9 (3): 176–185.

Facchin, Marco. 2024. "Neural Representations Unobserved—or: A Dilemma for the Cognitive Neuroscience Revolution." *Synthese* 203 (7): 1–42.

Farrell, Simon, and Stephan Lewandowsky. 2018. *Computational Modeling of Cognition and Behavior*. Cambridge: Cambridge University Press.

Fisher, Robert, Simon Perkins, Ashley Walker, and Erik Wolfart. 2003. "Laplacian/Laplacian of Gaussian." https://homepages.inf.ed.ac.uk/rbf/HIPR2/log.htm.

Fletcher, Samuel, Patricia Palacios, Laura Ruetsche, and Elay Shech. 2019. "Infinite Idealizations in Science: An Introduction." *Synthese* 196:1657–1669.

Fodor, Jerry. 1975. *The Modularity of Mind*. Cambridge, MA: MIT Press.

Frege, Gottlob. 1979. "Logic." In *Posthumous Writings*. Edited by Hans Hermes, Friedrich Kambartel, and Friedrich Kaulbach, 1–9. Translated by Peter Long and Roger White. Chicago: Chicago University Press.

Frigg, Roman. 2010. "Models and Fiction." *Synthese* 172:251–268.

Frigg, Roman, and James Nguyen. 2021. "Mirrors without Warning." *Synthese* 198:2427–2447.

Friston, Karl. 2010. "The Free-Energy Principle: A Unified Brain Theory." *Nature Reviews Neuroscience* 11:127–138.

Friston, Karl. 2013. "Life as We Know It." *Journal of the Royal Society Interface* 10:20130475.

Friston, Karl. 2019. "A Free Energy Principle for a Particular Physics." Unpublished manuscript.

Friston, Karl, Spyridon Samothrakis, and Read Montague. 2012. "Active Inference and Agency: Optimal Control without Cost Functions." *Biological Cybernetics* 106 (8): 523–541.

Friston, Karl, and Klaas Stephan. 2007. "Free-Energy and the Brain." *Synthese* 159 (3): 417–458.

Garden, Derek, Paul Dodson, Cian O'Donnell, Melanie White, and Matthew Nolan. 2008. "Tuning of Synaptic Integration in the Medial Entorhinal Cortex to the Organization of Grid Cell Firing Fields." *Neuron* 60 (5): 875–889.

Gardner, Martin. 1970. "Mathematical Games: The Fantastic Combinations of John Conway's New Solitaire Game 'Life.'" *Scientific American* 223 (4): 120–123.

Giere, Ronald. 1988. *Explaining Science: A Cognitive Approach.* Chicago: University of Chicago Press.

Giere, Ronald. 2006. *Scientific Perspectivism.* Chicago: University of Chicago Press.

Gładziejewski, Paweł. 2015. "Explaining Cognitive Phenomena with Internal Representations: A Mechanistic Perspective." *Studies in Logic, Grammar and Rhetoric* 40 (53): 63–90.

Gładziejewski, Paweł. 2016. "Predictive Coding and Representationalism." *Synthese* 193:559–582.

Gładziejewski, Paweł, and Marcin Milkowski. 2017. "Structural Representations: Causally Relevant and Different from Detectors." *Biology & Philosophy* 32:337–355.

Godfrey-Smith, Peter. 2003. *Theory and Reality.* Chicago: University of Chicago Press.

Godfrey-Smith, Peter. 2004. "On Folk Psychology and Mental Representation." In *Representation in Mind: New Approaches to Mental Representation,* edited by Hugh Clapin, Philip Staines, and Peter Slezak, 147–162. Amsterdam: Elsevier.

Godfrey-Smith, Peter. 2006. "Mental Representation, Naturalism, and Teleosemantics." In *Teleosemantics: New Philosophical Essays,* edited by David Papineau and Graham Macdonald, 42–68. Oxford: Clarendon Press.

Godfrey-Smith, Peter. 2009. "Models and Fictions in Science." *Philosophical Studies* 143:101–116.

Goodman, Nelson. 1976. *Languages of Art.* 2nd ed. Indianapolis: Hackett.

Grace, Lindsay. 2021. *Models of the Mind.* London: Bloomsbury.

Griffiths, Thomas, Dennis Norris, Nick Chater, and Alexandre Pouget. 2012. "How the Bayesians Got Their Beliefs (and What Those Beliefs Actually Are): Comment on Bowers and Davis." *Psychological Bulletin* 138 (3): 415–422.

Grush, Rick. 2004. "The Emulation Theory of Representation: Motor Control, Imagery, and Perception." *Behavioral and Brain Sciences* 27:377–396.

Häusser, Michael. 2000. "The Hodgkin-Huxley Theory of the Action Potential." *Nature Neuroscience* 3:1165.

Hesp, Casper, Max Ramstead, Axel Constant, Paul Badcock, Michael Kirchhoff, and Karl Friston. 2019. "A Multi-Scale View of the Emergent Complexity of Life: A Free Energy Proposal." In *Evolution, Development, and Complexity: Multiscale Models in Complex Adaptive Systems*, edited by Georgi Yordanov Georgiev, John M. Smart, Claudio L. Flores Martinez, and Michael E. Price, 195–227. Cham: Springer.

Hochstein, Eric. 2023. "Integration without Integrated Models or Theories." *Synthese* 202 (76): 1–25.

Hodgkin, Alan, and Andrew Huxley. 1952. "A Quantitative Description of Membrane Current and Its Application to Conduction and Excitation in Nerve." *Journal of Physiology* 117:500–544.

Hohwy, Jakob. 2012. "Attention and Conscious Perception in the Hypothesis Testing Brain." *Frontiers in Psychology* 3:1–14.

Hohwy, Jakob. 2016. "The Self-Evidencing Brain." Noûs 50 (2): 259–285.

Hohwy, Jakob. 2020. "Self-Supervision, Normativity and the Free Energy Principle." *Synthese* 199:29–53.

Hryvniak, David, Robert P. Wilder, Jeffrey Jenkins, and Siobhan M. Statuta. 2021. "Therapeutic Exercise." In *Braddom's Physical Medicine and Rehabilitation*, edited by David X. Cifu, 291–315. Philadelphia, PA: Elsevier.

Huneman, Paul. 2010. "Topological Explanations and Robustness in Biological Sciences." *Synthese* 177:213–245.

Hutto, Daniel. 2005. "Knowing What? Radical versus Conservative Enactivism." *Phenomenology and the Cognitive Sciences* 4:389–405.

Hutto, Daniel, and Erik Myin. 2013. *Radical Enactivism*. Cambridge, MA: MIT Press.

Isaac, Alistair. 2012. "Objective Similarity and Mental Representation." *Australasian Journal of Philosophy* 91:683–704.

Kadanoff, Leo. 2000. *Statistical Physics: Statics, Dynamics and Renormalization*. Singapore: World Scientific.

Kaplan, David. 2011. "Explanation and Description in Computational Neuroscience." *Synthese* 183:339–373.

Kellert, Stephen, Helen Longino, and C. Kenneth Waters. 2006. "Introduction: The Pluralist Stance." In *Scientific Pluralism*, edited by Stephen Kellert, Helen Longino, and C. Kenneth Waters, vii–xxix. Minneapolis: University of Minnesota Press.

Khalifa, Karim. 2020. "Understanding, Truth, and Epistemic Goals." *Philosophy of Science* 87:944–956.

Kiefer, Alex. 2017. "Literal Perceptual Inference." In *Philosophy and Predictive Processing*, edited by Thomas Metzinger and Wanja Wiese, 1–19. Frankfurt am Main: MIND Group.

Kiefer, Alex, and Jakob Hohwy. 2018. "Content and Misrepresentation in Hierarchical Generative Models." *Synthese* 195 (6): 1–29.

Kirchhoff, Michael. 2015. "Species of Realization and the Free Energy Principle." *Australasian Journal of Philosophy* 93 (4): 706–723.

Kirchhoff, Michael. 2018a. "Autopoiesis, Free Energy, and the Life-Mind Continuity Thesis." *Synthese* 195:2519–2540.

Kirchhoff, Michael. 2018b. "Hierarchical Markov Blankets and Adaptive Active Inference." *Physics of Life Reviews* 24: 1–2. https://doi.org/10.1016/j.plrev.2017.09.001.

Kirchhoff, Michael, and Julian Kiverstein. 2019. *Predictive Processing and Extended Consciousness: A Third-Wave View*. London: Routledge.

Kirchhoff, Michael, Julian Kiverstein, and Ian Robertson. Forthcoming. "The Literalist Fallacy and the Free Energy Principle: Model-Building, Scientific Realism and Instrumentalism." *British Journal for the Philosophy of Science*. https://doi.org/10.1086/720861.

Kirchhoff, Michael, Thomas Parr, Ensor Palacios, Karl Friston, and Julian Kiverstein. 2018. "The Markov Blankets of Life: Autonomy, Active Inference and the Free Energy Principle." *Journal of the Royal Society Interface* 15 (138). https://royalsocietypublishing.org/doi/epdf/10.1098/rsif.2017.0792.

Kirchhoff, Michael, and Ian Robertson. 2018. "Enactivism and Predictive Processing: A Non-Representational View." *Philosophical Explorations* 21 (2): 264–281.

Kitcher, Philip. 1991. "The Division of Cognitive Labor." *Journal of Philosophy* 87:5–22.

Klein, Colin. 2018. "What Do Predictive Coders Want?" *Synthese* 195:2541–2557.

Koch, Christof. 1999. *Biophysics of Computation*. New York: Oxford University Press.

Korzybski, Alfred. 1933. *Science and Sanity: An Introduction to Non-Aristotelian Systems and General Semantics*. Brooklyn: Institute of General Semantics.

Kriegeskorte, Nikolaus, and Pamela Douglas. 2018. "Cognitive Computational Neuroscience." *Nature Neuroscience* 21:1148–1160.

Kroon, Frederick. 2011. "Fictionalism in Metaphysics." *Philosophy Compass* 6 (11): 786–803.

Kroon, Frederick, Stuart Brock, and Jonathan McKeown-Green. 2019. *A Critical Introduction to Fictionalism*. London: Bloomsbury.

Laland, Kevin, John Odling-Smee, William Hoppitt, and Tobias Uller. 2013. "More on How and Why: Cause and Effect in Biology Revisited." *Biology & Philosophy* 28 (5): 719–745.

Laudan, Larry. 1977. *Progress and Its Problems: Towards a Theory of Scientific Growth*. Berkeley: University of California Press.

Lawler, Insa. 2021. "Scientific Understanding and Felicitous Legitimate Falsehoods." *Synthese* 198:6859–6887.

Lee, Jonny, and Daniel Calder. 2023. "The Many Problems with S-Representation (and How to Solve Them)." *Philosophy and the Mind Sciences* 4 (8): 1–31.

Levins, Richard. 1966. "The Strategy of Model Building in Population Biology." *American Naturalist* 54 (4): 421–431.

Levy, Arnon. 2011. "Information in Biology: A Fictionalist Account." Noûs 45 (4): 640–657.

Levy. 2015. "Modeling without Models." *Philosophical Studies* 172:781–798.

Mach, Ernest. 1911. *History and Root of the Principle of the Conservation of Energy*. Chicago: Open Court.

Maddy, Penelope. 1992. "Indispensability and Practice." *Journal of Philosophy* 89 (6): 275–289.

Mäki, Uskali. 2001. "Explanatory Unification: Double and Doubtful." *Philosophy of the Social Sciences* 31 (4): 488–506. https://doi.org/10.1177/004839310103100402.

Mäki. 2020. "Puzzled by Idealizations and Understanding Their Function." *Philosophy of the Social Sciences* 50 (3): 215–237.

Maley, Corey. 2023. "Analogue Computation and Representation." *British Journal for the Philosophy of Science* 74 (3): 736–769.

Maley, Corey, and Gualtiero Piccinini. 2014. "Neural Representation and Computation." In *Handbook of Neuroethics*, edited by Jens Clausen and Neil Levy, 79–94. Dordrecht: Springer.

Mann, Stephen, Ross Pain, and Michael Kirchhoff. 2022. "Free Energy: A User's Guide." *Biology & Philosophy* 37 (33): 1–35.

Marr, David. 1982. *Vision: A Computational Investigation into the Human Representation and Processing of Visual Information*. San Francisco, CA: Freeman.

Massimi, Michela. 2018. "Perspectival Modeling." *Philosophy of Science* 85 (3): 335–359.

Massimi, Michela. 2019. "Getting It Right." *Aeon*, January 28, 2019. https://aeon.co/essays/its-time-for-a-robust-philosophical-defence-of-truth-in-science.

Massimi, Michela. 2022. *Perspectival Realism*. Oxford: Oxford University Press.

Mayr, Ernst. 1961. "Cause and Effect in Biology." *Science* 134:1501–1506.

Maxwell, James Clerk. 1865. "VIII. A Dynamical Theory of the Electromagnetic Field." *Philosophical Transactions of the Royal Society of London* 155 (155): 459–512.

McClelland, John. 2013. "Integrating Probabilistic Models of Perception and Interactive Neural Networks: A Historical and Tutorial Review." *Frontiers in Psychology* 4. https://doi.org/10.3389/fpsyg.2013.00503.

McCulloch, Warren. 1974. "Recollections of the Many Sources of Cybernetics." *ASC Forum* 4 (2): 5–16.

McCulloch, Warren, and Walter Pitts. 1943. "A Logical Calculus of the Ideas Immanent in Nervous Activity." *Bulletin of Mathematical Biophysics* 7:115–133.

McMullin, Ernan. 1985. "Galilean Idealisation." *Studies in the History and Philosophy of Science* 16 (3): 247–273.

Milkowski, Marcin, and Mateusz Hohol. 2021. "Explanations in Cognitive Science: Unification versus Pluralism." *Synthese* 199:1–17.

Miller, Alexander. 2003. "The Significance of Semantic Realism." *Synthese* 136:191–217.

Millikan, R. G. 1984. *Language, Thought, and Other Biological Categories*. Cambridge, MA: MIT Press.

Millikan, Ruth. 1989. "Biosemantics." *Journal of Philosophy* 86:281–297.

Millikan, Ruth. 2000. *On Clear and Confused Ideas*. Cambridge: Cambridge University Press.

Mirza, Berk, Rick Adams, Christoph Mathys, and Karl Friston. 2016. "Scene Construction, Visual Foraging, and Active Inference." *Frontiers in Computational Neuroscience* 10. https://www.frontiersin.org/journals/computational-neuroscience/articles/10.3389/fncom.2016.00056/full.

Mitchell, Sandra. 2002. "Integrative Pluralism." *Biology & Philosophy* 17:55–70.

Morrison, Margaret. 2011. "One Phenomenon, Many Models: Inconsistency and Complementarity." *Studies in History and Philosophy of Science* 42:342–351.

Murphy, Kevin. 2012. *Machine Learning: A Probabilistic Perspective*. Cambridge, MA: MIT Press.

Neander, Karen. 2017. *A Mark of the Mental: A Defense of Informational Teleosemantics*. Cambridge, MA: MIT Press.

Newell, Alan. 1982. "The Knowledge Level." *Artificial Intelligence* 18:87–127.

Nguyen, James, and Roman Frigg. 2022. *Scientific Representation*. Cambridge: Cambridge Elements.

O'Brien, Gerard, and John Opie. 2004. "Notes towards a Structuralist Theory of Mental Representation." In *Representation in Mind: New Approaches to Mental Representation*, edited by Hugh Clapin, Philip Staines, and Peter Slezak, 1–20. Amsterdam: Elsevier.

O'Keefe, John, and Neil Burgess. 1978. *The Hippocampus as a Cognitive Map*. Oxford: Clarendon Press.

O'Keefe, John, and Neil Burgess. 1996. "Geometric Determinants of the Place Fields of Hippocampal Neurons." *Nature* 381:425–428.

O'Keefe, John, and Neil Burgess. 2005. "Dual Phase and Rate Coding in Hippocampal Place Cells: Theoretical Significance and Relationship to Entorhinal Grid Cells." *Hippocampus* 15:853–866.

Orlandi, Nico. 2020. "Representing as Coordinating with Absence." In *What Are Mental Representations?*, edited by Joulia Smortchkova, Krys Dolega, and Tobias Schlicht, 101–134. New York: Oxford University Press.

Pain, Ross, Michael Kirchhoff, and Stephen Mann. 2022. "Life Has 'Surprises': Can an Ambitious Theory Unify Biology, Neuroscience and Psychology?" *The Conversation*, August 14, 2022. https://theconversation.com/life-hates-surprises-can-an-ambitious-theory-unify-biology-neuroscience-and-psychology-184052.

Palacios, Ensor, Adel Razi, Thomas Parr, Michael Kirchhoff, and Karl Friston. 2020. "On Markov Blankets and Hierarchical Self-Organization." *Journal of Theoretical Biology* 486:110089.

Panagiotatou, Maria, and Statis Psillos. 2023. "Review of 'Perspectival Realism.'" *Notre Dame Philosophical Reviews*, July 5, 2023. https://ndpr.nd.edu/reviews/perspectival-realism/.

Papineau, David. 2010. "Realism, Ramsey Sentences and the Pessimistic Meta-Induction." *Studies in History and Philosophy of Science* 41:375–385.

Parr, Thomas, and Karl Friston. 2018. "The Anatomy of Inference: Generative Models and Brain Structure." *Frontiers in Computational Neuroscience* 12: https://doi.org/10.3389/fncom.2018.00090.

Parr, Thomas, Berk Mirza, Hayriye Cagnan, and Karl Friston. 2019. "Dynamic Causal Modelling of Active Inference." *Journal of Neuroscience* 39 (32): 6265–6275.

Parr, Thomas, Giovanni Pezzullo, and Karl Friston. 2022. *Active Inference*. Cambridge, MA: MIT Press.

Prandtl, Ludwig. 1904. "Motion of Fluids with Very Little Viscosity." *German History Intersections*. https://germanhistory-intersections.org/en/knowledge-and-education/ghis:document-202.

Pearl, Judea. 1988. *Probabilistic Reasoning in Intelligent Systems: Networks of Plausible Inference*. San Francisco: Morgan Kaufmann.

Piccinini, Gualtiero. 2015. *Physical Computation: A Mechanistic Account*. New York: Oxford University Press.

Piccinini, Gualtiero. 2020. *Neurocognitive Mechanisms: Explaining Biological Cognition*. New York: Oxford University Press.

Piccinini, Gualtiero. 2022. "Situated Neural Representations: Solving the Problems of Content." *Frontiers in Neurorobotics* 16:846979.

Piccinini, Gualtiero, and Sonya Bahar. 2013. "Neural Computation and the Computational Theory of Cognition." *Cognitive Science* 34:453–488.

Piccinini, Gualtiero, and Eric Thompson. 2018. "Neural Representations Observed." *Minds and Machines* 28:191–235.

Planer, Ron, and Peter Godfrey-Smith. 2021. "Communication and Representation Understood as Sender-Receiver Coordination." *Mind & Language* 36:750–770.

Poincaré, Henry. 1952. *Science and Hypothesis*. New York: Dover.

Potochnik, Angela. 2017. *Idealization and the Aims of Science*. Chicago: Chicago University Press.

Potochnik, Angela. 2023. "Truth and Reality: How to be a Scientific Realist without Believing Scientific Theories Should Be True." In *Scientific Understanding and Representation: Modeling in the Physical Sciences*, edited by Insa Lawler, Karim Khalifa, and Elay Shech, 1–24. New York: Routledge.

Putnam, Hilary. 1975. *Philosophical Papers*. Volume 1, *Mathematics, Matter and Method*. Cambridge: Cambridge University Press.

Putnam, Hilary. 1978. *Meaning and the Moral Sciences*. London: Routledge and Kegan Paul.

Psillos, Statis. 1999. *Scientific Realism: How Science Tracks Truth*. London: Routledge.

Psillos, Statis. 2011. "Living with the Abstract: Realism and Models." *Synthese* 180 (1): 3–17.

Ptolemy, Claudius. 1984. *Ptolemy's Almagest*. Translated and annotated by G. J. Toomer. London: Duck.

Pylyshyn, Zenon. 1984. *Computation and Cognition*. Cambridge, MA: MIT Press.

Quine, Willard Van Orman. 1969. "Natural Kinds." In *Ontological Relativity and Other Essays*, 114–138. New York: Columbia University Press.

Ramsey, William. 2007. *Representation Reconsidered*. Cambridge: Cambridge University Press.

Ramsey, William. 2018. "Maps, Models and Computational Simulations in the Mind." In *The Routledge Handbook of the Computational Mind*, edited by Mark Sprevak and Matteo Colombo, 259–271. London: Routledge.

Ramsey, William. 2020. "Defending Representational Realism." In *What Are Mental Representations?*, edited by Joulia Smortchkova, Krys Dolega, and Tobias Schlicht, 54–78. New York: Oxford University Press.

Ramsey, William. 2023. "The Hard Problem of Content Is Neither." *Review of Philosophy and Psychology*: 1–22. https://doi.org/10.1007/s13164-023-00714-9.

Ramstead, Max, Paul Badcock, and Karl Friston. 2019. "Variational Neuroethology: Answering Further Questions: Reply to Comments on 'Answering Schrödinger's Question: A Free-Energy Formulation.'" *Physics of Life Reviews* 24:59–66.

Ramstead, Max, Karl Friston, and Ines Hipólito. 2020. "Is the Free-Energy Principle a Formal Theory of Semantics? From Variational Density Dynamics to Neural and Phenotypic Representations." *Entropy* 22:889. doi:10.3390/e22080889.

Ramstead, Max, Michael Kirchhoff, Axel Constant, and Karl Friston. 2021. "Multiscale Integration: Beyond Internalism and Externalism." *Synthese* 98:41–70.

Rawls, John. 1971. *A Theory of Justice*. Cambridge, MA: Harvard University Press.

Rescorla, Michael. 2009. "Cognitive Maps and the Language of Thought." *British Journal for the Philosophy of Science* 60:377–407.

Rescorla, Michael. 2015. "Bayesian Perceptual Psychology." In *The Oxford Handbook of the Philosophy of Perception*, edited by Mohan Matthen, 694–716. Oxford: Oxford University Press.

Rice, Colin. 2018. "Idealized Models, Holistic Distortions, and Universality." *Synthese* 195: 2795–2819.

Rice, Colin. 2019. "Models Don't Decompose That Way: A Holistic View of Idealized Models." *British Journal for the Philosophy of Science* 70:179–208.

Rice, Colin. 2021. "Understanding Realism." *Synthese* 198:4097–4121.

Robertson, Ian, Julian Kiverstein, and Michael Kirchhoff. 2024. "Is the Free Energy Principle for Real?" *British Journal for the Philosophy of Science* 76. https://www.thebsps.org/short-reads/is-the-free-energy-principle-for-real/.

Rowbottom, Darrell. 2019. *The Instrument of Science: Scientific Anti-Realism Revitalised*. New York: Routledge.

Rowbottom, Darrell. 2022. "Can Meaningless Statements Be Approximately True? On Relaxing the Semantic Conception of Scientific Realism." *Philosophy of Science* 89:879–888.

Schelling, Thomas. 1978. *Micromotives and Macrobehavior*. New York: Norton.

Schulte, Peter, and Karen Neander. 2022. "Teleological Theories of Mental Content." *Stanford Encyclopedia of Philosophy* (Summer 2022 Edition), edited by Edward N. Zalta. https://plato.stanford.edu/archives/sum2022/entries/content-teleological/.

Schwartenbeck, Philipp, Thomas FitzGerald, Chris Mathys, Ray Dolan, Friedrich Wurst, Martin Kronbichler, and Karl Friston. 2015. "Optimal Inference with Suboptimal Models: Addiction and Active Bayesian Inference." *Medical Hypotheses* 84 (2): 109–117.

Sejnowski, Terrence, Paul Churchland, and Cristof Koch. 1988. "Computational Neuroscience." *Science* 241:1299–1306.

Sganzerla, Lorena, Dan Hutto, and Michael Kirchhoff. Forthcoming. "Getting Beyond the Extended Mind: New Argument for Extensive Enactivism." *Synthese*.

Shagrir, Oron. 2012. "Structural Representations and the Brain." *British Journal for the Philosophy of Science* 63:519–545.

Shagrir, Oron. 2022. *The Nature of Physical Computation*. Oxford: Oxford University Press.

Shapiro, Lawrance. 2017. "Mechanism or Bust? Explanation in Psychology." *British Journal for the Philosophy of Science* 68:1037–1059.

Shea, Nick. 2013. "Naturalizing Representational Content." *Philosophy Compass* 8 (5): 496–509.

Shea, Nick. 2020. *Representation in Cognitive Science*. New York: Oxford University Press.

Shech, Elay. 2018. "Infinite Idealization in Physics." *Philosophy Compass* 13:1–15.

Shech, Elay. 2023. *Idealizations in Physics*. Cambridge: Cambridge University Press.

Spivey, Michael. 2007. *The Continuity of Mind*. New York: Oxford University Press.

Sprevak, Mark. 2010. "Computation, Individuation, and the Received View on Representation." *Studies in History and Philosophy of Science* 41:260–270.

Sprevak, Mark. 2013. "Fictionalism about Neural Representation." *The Monist* 96:539–560.

Sprevak, Mark. 2018. "Realism about Cognitive Science." In *The Routledge Handbook of Scientific Realism*, edited by Juha Saatsi, 357–368. London: Routledge.

Sprevak, Mark, and Matteo Colombo, eds. 2018. *The Routledge Handbook of the Computational Mind*. London: Routledge.

Sterelny, Kim. 1991. *The Representational Theory of Mind: An Introduction*. Cambridge MA: Blackwell.

Sterratt, David, Bruce Graham, Andrew Gillies, and David Wilshaw. 2011. *Principles of Computational Modelling in Neuroscience*. Cambridge: Cambridge University Press.

Stinson, Catherine. 2020. "From Implausible Artificial Neurons to Idealized Cognitive Models: Rebooting Philosophy of Artificial Intelligence." *Philosophy of Science* 87:590–611.

Strevens, Michael. 2008. *Depth: An Account of Scientific Explanation*. Cambridge, MA: Harvard University Press.

Strevens, Michael. 2017. "How Idealizations Provide Understanding." In *Explaining Understanding: New Perspectives from Epistemology and Philosophy of Science*, edited by Stephen Grimm, Christoph Baumberger, and Sabine Ammom, 1–13. London: Routledge.

Strevens, Michael. 2019. "The Structure of Asymptotic Idealization." *Synthese* 196:1713–1731.

Suárez, Mauricio. 2003. "Scientific Representation: Against Similarity and Isomorphism." *International Studies in the Philosophy of Science* 17 (3): 225–244.

Suárez, Mauricio. 2010. "Fictions, Inference, and Realism." In *Fictions and Models: New Essays*, edited by John Woods, 225–246. Munich: Philosophia Verlag.

Sun, Ron, ed. 2023. *The Cambridge Handbook of Computational Cognitive Sciences*. Cambridge: Cambridge University Press.

Swoyer, Chris. 1991. "Structural Representation and Surrogative Reasoning." *Synthese* 87:449–508.

Thompson-Jones, Martin. 2005. "Idealization and Abstraction: A Framework." In *Idealization XII: Correcting the Model*, edited by Martin Thompson-Jones and Nancy Cartwright, 173–217. Amsterdam: Rodopi.

Thompson-Jones, Martin. 2020. "Realism about Missing Systems." In *The Scientific Imagination: Philosophical and Psychological Perspectives*, edited by Arnon Levy and Peter Godfrey-Smith, 75–101. New York: Oxford University Press.

Toon, Adam. 2023. *Mind as Metaphor: A Defence of Mental Fictionalism*. Oxford: Oxford University Press.

Turing, Alan. 1937. "On Computable Numbers, with an Application to the Entscheidungsproblem." *Proceedings of the London Mathematical Society* s2-42 (1): 230–265.

Turner, Brandon, and Trisha van Zandt. 2018. "Approximating Bayesian Inference through Model Simulation." *Trends in Cognitive Sciences* 22 (9): 1–15.

Tversky, Amos. 1977. "Features of Similarity." *Psychological Review* 84:327–352.

Vaihinger, Hans. 1924. *The Philosophy of "As-If"*. Translated by C. K. Ogden. New York: Harcourt Brace.

Van Es, Thomas. 2021. "Living Models or Life Modelled? On the Use of Models in the Free Energy Principle." *Adaptive Behavior* 29 (3): 315–329.

Van Fraassen, Bas. 1980. *The Scientific Image*. Oxford: Oxford University Press.

Van Gelder, Tim. 1995. "What Might Cognition Be, if not Computation?" *Journal of Philosophy* 91 (7): 345–381.

Wallace, Meg. 2016. "Saving Mental Fictionalism from Cognitive Collapse." *Res Philosophica* 93 (2): 405–424.

Wallis, Charles. 1994. "Representation and the Imperfect Ideal." *Philosophy of Science* 61:407–428.

Walmsley, Lachlan. 2021. "The Strategy of Model Building in Climate Science." *Synthese* 199:745–765.

Walton, Kendall. 1990. *Mimesis as Make-Believe: On the Foundations of the Representational Arts*. Cambridge, MA: Harvard University Press.

Weisberg, Michael. 2006. "Forty Years of 'The Strategy': Levins on Model Building and Idealization." *Biology & Philosophy* 21:623–645.

Weisberg, Michael. 2007. "Three Kinds of Idealization." *Journal of Philosophy* 104 (12): 639–659.

Weisberg, Michael. 2012. "Getting Serious about Similarity." *Philosophy of Science* 79:785–794.

Weisberg, Michael. 2013. *Simulation and Similarity: Using Models to Understand the World*. New York: Oxford University Press.

Weisberg, Michael. 2018. "Validating Idealized Models." In *The Experimental Side of Modeling*, edited by Isabelle Peschard and Bas van Fraassen, 240–264. Minneapolis: University of Minnesota Press.

Weiskopf, Daniel. 2011. "Models and Mechanisms in Psychological Explanation." *Synthese* 183:313–338.

Williamson, Timothy. 2017. "Model-Building in Philosophy." In *Philosophy's Future: The Problems of Philosophical Progress*, edited by Russell Blackford and Damien Broderick, 159–171. London: John Wiley and Sons.

Wimsatt, William. 1987. "False Models as Means to Truer Theories." In *Neutral Models in Biology*, edited by Matthew Nitecki and Antoni Hoffman, 23–55. New York: Oxford University Press.

Wimsatt, William. 2007. *Re-Engineering Philosophy for Limited Beings: Piecewise Approximations to Reality*. Cambridge, MA: Harvard University Press.

Woodward, James. 2003. *Making Things Happen: A Theory of Causal Explanation.* Oxford: Oxford University Press.

Yablo, Stephen. 2014. *Aboutness*. Princeton, NJ: Princeton University Press.

Yablo, Stephen. 2020. "Models and Reality." In *The Scientific Imagination: Philosophical and Psychological Perspectives*, edited by Arnon Levy and Peter Godfrey-Smith, 128–153. New York: Oxford University Press.

Young, Hugh, and Roger Freedman. 2007. *University Physics with Modern Physics.* 12th ed. Reading, MA: Addison-Wesley.

Zhang, Ge, Xiaoqiang Yan, Yulong Xu, and Yangdong Ye. 2020. "Neural Guided Visual Slam System with Laplacian of Gaussian Operator." *IET Computer Vision* 15 (3): 181–196.

Index

Abstraction, 31, 36–39, 79, 92–93, 108, 120, 142, 163, 198
Andrews, Mel, 136, 191
Appiah, Anthony K., 40
Approximate Bayesian inference, 190–191, 194
Approximation, 51, 118–120, 153, 189–191

BART system, 58–59
Batterman, Robert, 44, 78, 182
Bayesian cognitive science, 4–7, 17–18, 83, 134–136, 162–169, 182–196
Bayes' rule, 134–135, 144, 190–194
Borges, Jorge L., 35
Bruineberg, Jelle, 185–189

Calcott, Brett, 151–152
Canonical neural computation, 141–144
Cao, Rosa, 88, 92, 104, 107–111, 120–123
Cartwright, Nancy, 53, 158
Chakravartty, Anjan, 48, 67, 75, 120
Chalmers, David, 14, 145
Chirimuuta, Mazviita, 71, 79, 88, 107–113, 124, 138, 140–141, 150, 154–155, 161–169
Clark, Andy, 83
Coelho Mollo, Dimitri, 89, 100–102, 106, 129, 150, 154, 160, 163
Combinatorial state-automata, 145
Complexity, 1, 11–13, 26, 102, 153–156
Computational modeling, 13, 92, 127–196
Computational theory of cognition, 13, 19, 92, 128–130, 136, 202
Connectionism, 24, 154, 162, 166–169
Covariation, 7, 98–99, 115–116, 193
Craver, Carl, 26, 142

DEKI, 61–63
Dennett, Daniel, 121, 127–128, 204–205
Denotation, 56–57, 59
Dretske, Fred, 98–99, 129
Dynamical systems theory, 24, 154
Dynamicism, 24

Efficient coding models, 140–141, 169, 190
Egan, Frances, 14–15, 88, 91, 106, 119, 121, 127–136, 144–145
Elgin, Catherine, 32, 40–41, 50–51, 56–57, 72–75, 165
Eliasmith, Chris, 24
Eliminativism, 6, 109, 129, 207
Entity realism, 6, 11, 15, 105, 123, 125, 216
Entropy, 30, 81, 115, 141–142, 169, 172, 178, 181–182, 191–196, 208–209, 212

Explanation, 3, 12, 15, 23, 32–33, 42–43, 49–63, 68–84, 89–90, 108, 111–113, 123–126, 130, 132, 140, 171
explanatory pluralism, 15–17, 149–150, 166–169
consistency-friendly pluralism, 16–17, 165–169
inconsistent-friendly pluralism, 16–17, 155, 162–164
explanatory unification, 2–4, 7, 9, 16–19, 83, 93–95, 149–157, 169, 172, 197, 215
function-theoretic explanation, 133–136
mechanistic explanation, 137–143, 168
decomposition, 139–142
localization, 139–142
proximate causes, 151–152
ultimate causes, 151–152

Facchin, Marco, 106, 116–117
Fictionalism, 6–7, 19–20, 31–32, 55, 68, 87, 98, 127, 133, 181–184, 197–216
fictional models/systems, 198
fictions, 31, 62, 73, 88, 98, 174, 183–185, 197–216
mental fictionalism, 55, 98, 202–204
neural computation fictionalism, 208
neural representation fictionalism, 207–208
Fodor, Jerry, 13, 138
Folk psychology, 127, 203–204
Free energy principle, 17–18, 83, 94, 169, 171–196
Bayesian inference, 182, 189–191, 194
expected free energy, 173
generative model, 167, 175, 177, 189–194
grand unifying theory, 17, 94, 149, 171–196
Jensen's inequality, 179, 205–206
Markov blanket formalism, 175, 185, 189
variational free energy, 81, 173–183, 191, 203–205, 208–209, 212
Frege, Gottlob, 83
Frigg, Roman, 9, 28–33, 48–49, 56, 61–63, 68, 74–75, 189, 202–203
Friston, Karl, 18, 140, 142, 171–176, 179, 181–182, 184, 187–188, 190, 193

Galileo, Galilei, 124, 179, 203
Godfrey-Smith, Peter, 2, 25, 30, 32–33, 40, 65, 81–82, 89, 99, 101, 109, 183, 198, 202, 210

Hochstein, Eric, 16, 150, 156, 159–160, 163, 166
Hohwy, Jakob, 171, 181, 191–192, 194–196
Hutto, Daniel, 7, 89, 91, 106–107, 111, 115–116, 215
Hypotheses, 198–199

Idealization, 1–28, 31–32, 34, 36–45, 47–63, 65–78, 80–84, 87–95, 97–111, 113–139, 141–161, 163–213
deliberate distortions, 2, 31, 47, 50, 54
falsehood view of idealization (FvI), 8, 40–53, 56–72, 125, 128, 178, 180, 199, 212
Galilean idealization, 42–43, 124–125, 152–153, 177–179
imaginary systems, 2, 53, 62, 189, 203
legitimate falsehoods, 2, 49, 63, 67
misrepresentation, 2, 8, 37, 40–45, 48, 77, 99, 101, 213
multiple-models idealization (MMI), 15, 17, 94, 149–169

nonliteral view of idealization, 17, 19–21, 47–63, 65–66, 69, 76–78, 81–83, 87, 89, 95, 104, 125, 128, 160, 171, 197, 199, 201–202, 204, 206, 208, 210, 212–213, 216
nonpermanent view of idealization, 42–43, 123–124
permanent view of idealization, 43, 123–124
Information processing, 3, 122, 138, 159, 192, 215
Intentional stance, 127, 204
Interpretation, 189–190

Kaplan, David, 41–42
Kiefer, Alex, 194
Klein, Colin, 177–179

Language of thought hypothesis, 6
Levins, Richard, 23
Levy, Arnon, 60, 71
Literalist fallacy, 75, 174
Literalism, 5, 8, 47–49, 60, 70, 72, 74–75, 83, 91–94, 164, 173–174, 185

Machine learning, 5, 15, 17, 91, 95, 141, 175, 178, 183, 190, 194
Mäki, Uskali, 44, 52, 54, 62, 93, 114
Maley, Corey, 3, 13, 137, 143
Map problem, 32–36, 76, 87, 95, 172
Marr, David, 130–131, 134–136, 161–162, 167–169
algorithmic level, 130
computational level, 130
implementational level, 130
Massimi, Michela, 150, 154–156, 160–166
McMullin, Ernan, 31, 178
Milkowski, Marcin, 149
Mitchell, Sandra, 150–151, 155, 158, 163
Model-based science, 1–2, 4, 7, 10, 15, 17, 20, 29, 36, 44, 60, 67–70, 74, 88, 90, 103, 196–198, 202, 204, 212, 215
model-building, 1, 116, 213
Morrison, Margaret, 150, 154–155, 157
Mutual information, 30, 193, 195

Neural computation, 2–7, 9–10, 83, 102, 141, 143, 144, 197, 215–216
Laplacian of a gaussian distribution, 15, 59, 79, 91, 134–136, 144, 208–209
mathematical content, 14–15, 128–136
mechanistic theory of computation, 14–15, 137–148, 168
rules, 2, 15, 26, 30, 137, 142, 144–145, 207
Neural representation, 2–7, 9–12, 71, 83, 88–90, 97–126, 133, 197, 202, 204, 207–209, 216
causal efficacy, 11, 101, 106
content determinacy, 11, 101
semantic content, 6, 101, 111, 115–116, 127–133
structural representation, 116–120, 201
vehicles, 6, 97, 101–103, 111, 119, 126, 132–147
Nguyen, James, 9, 28–29, 31, 33, 48–49, 56, 61–62, 68, 74–75, 189
Nomic covariation, 98–99. *See also* Dretske, Fred
Nonliteralism, 69–70, 165, 201

Papineau, David, 73
Piccinini, Gualtiero, 3, 13–15, 27–28, 52, 92–93, 102, 106, 116, 130, 137–147
Place cells, 103–104, 117, 119, 159, 207
Point objects, 2–3, 12, 53, 63, 89, 98, 104, 108, 123, 216
Potochnik, Angela, 1, 40–43, 53, 73, 75, 121, 149–156, 211

Prediction error minimization (PEM), 134, 192
Problem of inconsistent models (PIM), 17, 94, 150
Psillos, Statis, 33, 48, 66, 69–72, 150
  Psillos' central dilemma, 69–72

Quine, Willard V. O., 120

Ramsey, William, 7, 88, 102, 111, 116, 121, 128
Representational theory of mind (RTM), 6, 11, 19, 88, 97, 105, 126
Rescorla, Michael, 7, 116
Rowbottom, Darrell, 9, 33, 48, 59–60, 72, 82, 164

Scientific antirealism, 20, 33, 36, 68–69, 83
  instrumentalism, 69, 80–84, 95, 127, 173–174, 182–185, 190, 205
  pessimistic meta-induction, 80
Scientific fictionalism, 20, 210–214
Scientific model, 1–84, 110, 118, 136, 163, 173, 198–199, 202–203
  baseball model, 52–53
  billiard ball model of gasses, 198–199
  cannonball model, 38–39
  center of gravity, 204
  computational model, 3–4, 13–14, 18, 26, 29–35, 52, 79, 84, 92, 103, 127–136, 137–147, 159, 161, 190
  concrete model, 29
  game of life model, 32–33
  generalized model, 34
  Hardy-Weinberg model, 51–52, 136, 160, 203
  Hodgkin-Huxley (HH) model, 1, 23–25
  ideal gas model, 4, 11–12, 50, 54, 57, 60, 106, 191
  infinite system, 37
  limit function, 78–80
  linear harmonic oscillator, 70–71
  mathematical model, 25, 28–36, 78–79, 84, 92–95, 105, 133, 141, 157, 171–175, 193–196
  McCulloch and Pitts model, 27, 29, 31, 33
  network traffic model, 140
  Newton's laws of motion, 91
  ocular motor system model, 122
  Prandtl's model of fluid dynamics, 157
  Ptolemy's model of the solar system, 31
  Schelling model, 26–30, 32–33, 118–119, 203
  target-directed model, 34
  targetless model, 33
  weather models, 153
Scientific realism, 2, 4–20, 32–36, 48–49, 65–84, 89, 93, 95, 124–126, 132, 144–145, 150, 154–169, 173–174, 178, 182–189, 210–213
  deflationary scientific realism, 17, 150, 154–155, 160, 163
  epistemic stance, 66, 68–69, 84
  individual scientific realism, 80
  metaphysical stance, 66
  perspectival scientific realism, 17, 150, 154–155, 160–162, 164–166
  selective scientific realism, 80–81
  semantic stance, 66
Scientific representation, 61, 68, 120, 155, 184, 216
  p-representation, 56–57
  representation *of p*, 56–57, 61
Semantic realism, 4–8, 48, 72
  restrictive semantic realism, 66, 72, 82
  unrestrictive semantic realism, 66, 82 , 208
Shagrir, Oron, 12, 107, 116–126, 129–130, 138–142
Shannon entropy, 30, 193–195
Shea, Nick, 88–89, 99–100, 103–107, 111–116, 118–119, 123, 129, 201

Shech, Elay, 40, 68, 78
Similarity, 27, 39, 49, 67, 117–120, 129
Simplification, 13, 23–24, 28, 41, 51–52, 71, 90, 108, 113, 138, 153, 177–178
Spivey, Michael, 122
Sprevak, Mark, 94, 127, 129–133, 153–154, 201–209
Suárez, Mauricio, 68, 74, 180, 184–185, 202, 210, 212–213
State transitions, 14–15, 24–30, 34–35, 93, 122, 143–145, 192
  continuous, 28, 42
  discrete, 28, 42
Stinson, Catherine, 3
Strevens, Michael, 40–44, 50–51, 68, 76–78, 80, 106, 125, 139, 180
Structural correspondence, 104, 118–119, 201

Target system, 10, 15, 24, 31–36
Teleosemantics, 99
  representational consumer, 99–100, 106
  representational producer, 99–100
Thomas, van Es, 182
Thompson-Jones, Martin, 2, 31, 38–44, 63, 90
Toon, Adam, 32, 55, 98, 197–204
Tractability, 23, 54–55, 99, 92, 104, 106, 175
  cognitive tractability, 88, 107, 110, 113, 123
  computational tractability, 12, 49, 52, 54–55, 60, 92, 106–107, 123
Translation task, 104–107, 112–115, 120, 126, 132–133
Truth, 2, 5, 8, 24, 32–33, 54, 66–84, 105, 124, 150, 164, 178, 185. *See* Scientific realism
  approximate truth, 2, 35–36, 65–67, 71–79, 80–82, 128, 135–136, 149, 154, 161, 183, 189, 211
  partial truth, 60
Turing machine, 24, 34, 57, 153

Vaihinger, Hans, 198–199, 204
Veritism, 8, 67, 69, 72–77, 149, 154, 173–174, 211–212

Wallis, Charles, 99, 106
Weisberg, Michael, 2, 15, 23, 25, 27, 29–33, 40–42, 68, 94, 118, 124, 149, 152–154, 169, 178
Wimsatt, William, 41, 158

Publisher contact:
The MIT Press
Massachusetts Institute of Technology
77 Massachusetts Avenue, Cambridge, MA 02139
mitpress.mit.edu

EU Authorised Representative:
Easy Access System Europe, Mustamäe tee 50,
10621 Tallinn, Estonia
gpsr.requests@easproject.com

Printed by Integrated Books International,
United States of America